Electronics for Biologists

Fundamental Concepts

Edition 1.07

by Timothy J. Gawne

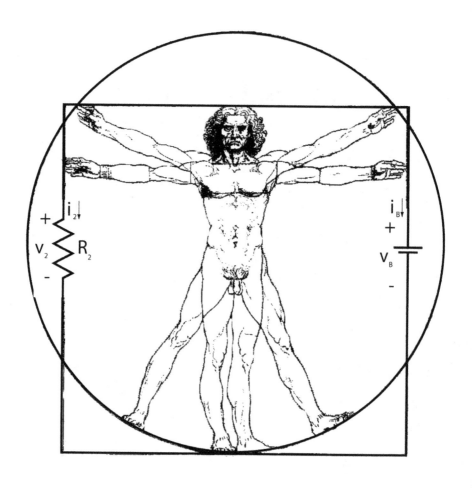

Published by: Ballacourage Books, Framingham, MA
Cover design by J. Cameron McClain
LCCN: 2001012345
ISBN: 978-0-9852956-0-8

Table of Contents

Forward

This is a text in the basics of electronics that is tailored for students who are not engineering or physics majors. If you are a student in the biological sciences, and you need a background in electronics, either because your field of study requires a working knowledge of electric circuit concepts (such as ion channels, electrophysiology, MRI imaging, etc.), your work requires the use of amplifiers and filters, or you want to be able to use computers to acquire data and control experiments, then this reference is for you.

This is not a substitute for a general electronics text. Many topics important to the study of electronics per se have been omitted to maintain a focus on topics of relevance to biologists. Be warned that we are skipping the derivations of equations, and more advanced topics are often covered only in an outline, with occasional advice to "consult with a real engineer".

There is little emphasis on circuit design, as most equipment that a biologist could want can be purchased for a modest amount. Remember the first rule of engineering: never re-invent the wheel! Rather, the basic principles are emphasized, as these are critical to an understanding of circuit models of biological processes and to the fundamental limitations of electrical instruments. Special emphasis is given to problems of particular biological relevance, such as recording neural signals from microelectrodes and circuit models of biological membranes. Additionally, the practices and pitfalls of such operations as filtering and shielding are covered in especial depth.

The author acknowledges the assistance of Huong Lan Vu and Tyler Dickerhoff in finding, and correcting, errors in the text.

Timothy J. Gawne
University of Alabama at Birmingham
January 2012

Chapter 1. Lumped Quasi-Static Models

It is important to remember what it is, exactly, that we mean by an electric circuit. The real physical world consists of electric and magnetic fields, electric charges, and materials with different electrical and magnetic properties. The interplay of these elements is defined over all three dimensions and time by Maxwell's equations. However, solving Maxwell's equations is very hard if not impossible for all but the simplest geometries.

Fortunately, for many purposes it suffices to use a simplified method of analysis that most of us are familiar with as standard electric circuit diagrams. These diagrams abstract and simplify the great complexity of the real world into models that, under the right conditions, can be very accurate indeed.

Consider the circuit in Figure 1-1. Panel A shows a battery hooked up to a resistor. The geometry is complex, and a full solution for the exact pattern of electric fields and currents is not a trivial task. Voltage varies continuously at all points, both through the wires and inside the different parts of the battery and resistor.

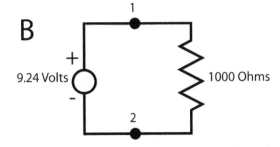

Figure 1-1. A. The real world. B. A lumped-element *model* of same.

Currents flow through the wires and components in complex patterns, much as water flow varies with position in a river, and some current leaks out of the wires to flow through the air and table-top. The circuit elements act as antennas, and so external electromagnetic sources like radio transmitters or electric lights affect the operation of the circuit. The currents flowing in the wires create external magnetic fields, which can interact with other objects. Different parts of the circuit that are not directly connected to each other can nonetheless affect each other via the electric field.

However, the simplified model circuit in panel B is easy to analyze. We say that this is a *lumped quasi-static model*. Lumped, because the full complexity of the 3D geometry is reduced to simple lumped elements connected by uni-dimensional nodes. Quasi-static because, even though such models can handle time-varying signals, they do not model the true dynamics of how electric and magnetic fields interact through space.

In panel A, voltage is a continuous function of space. In panel B, there are only two voltages: the voltage at node 1 and the voltage at node 2. In panel A, the resistor is a physical object with real size and where current and voltage can vary in complex ways inside it. The abstract resistor in panel B has no internal structure at all, and is completely specified by its value of 1000 ohms, and by the fact that it connects node 1 with node 2. The model resistor has no other properties.

Note that in the lumped circuit model the position of the elements, their sizes, or how many bends the wires have, does not matter. It is only the basic topology of the network that count, i.e., what connects to what and in what order. All other details are irrelevant.

For all its simplification, the model can often provide very accurate predictions of what the real circuit will do. When is the model good enough, and when does it fail to give a sufficiently accurate model of the world? This is a tough question with no precise answer. At the simplest level it depends upon how accurate a result you want, for the more accuracy you demand the more you have to consider all aspects of the physical reality of the system under study.

Very roughly, if the frequencies used are such that the wavelength of the signals in question becomes similar to the physical size of the network, things start to break down. With light traveling at about 300,000 km/sec in a vacuum, somewhat less in wires, signals in the audio range (up to about 20,000 cycles/sec) don't need to worry about this, but frequencies in the gigahertz (such as microwave) frequency often do.

Other cases where this model breaks down involve very fast transients, that is, rapid changes in voltage or current. In these cases you can get what are in effect "echoes" bouncing around in a circuit. As electric signals travel about a foot in a nanosecond, for any circuit that can fit into a room changes on the millisecond level obviously do not have to worry about these effects, because any "echoes" back and forth through the circuit will decay before you can notice them. We will discuss these effects briefly in a later chapter, for now, just remember that for slowly varying signals – such as those in the audio range, or the signals that come from most biological processes – these effects are usually negligible.

Also, when dealing with external electric or magnetic fields, or very weak signals, it is often the case that the physical geometry of the circuit must be taken into account. In effect, the circuit can act as an antenna, and in this case the specific geometry and orientation of the circuit can become important. Note that standard electronic design techniques have evolved to specifically limit such effects. Also, even when the simple circuit diagrams break down, it is often possible to model this breakdown by simply adding more idealized circuit elements to model the effects of (for example) external noise.

So powerful are the techniques of circuit analysis, that even when a problem specifically requires a solution of the full three-dimensional pattern of current in a substance, it is often easier to just subdivide the substance into small lumped "chunks", and model the connections between "chunks" with standard lumped circuit models. This technique is known as *Finite Element Analysis* (see Figure 1-2). The finite elements obviously can not completely reproduce the behavior of currents and voltages in materials that are continuously curved, but by increasing the number of finite elements used to model the real system you can reduce the error to any desired level.

Finite element analysis is therefore a full circle, where a technique of analysis designed to avoid the full complexity of electromagnetic problems in three dimensions is turned back and used to solve the same problems whose development it was to avoid in the first place...

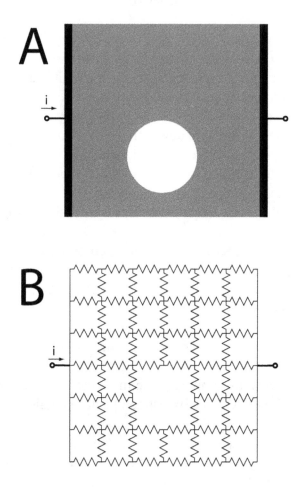

Figure 1-2. A. An electric current is made to flow across a partially conductive sheet with a hole in it. B. Finite-element model of the above.

Chapter 2. Fundamental Concepts

Electrical Charge

Most of electronics concerns the motion of *electric charge*. Charge exists in two varieties, arbitrarily labeled positive and negative. Opposite charges attract and like charges repel. Thus, most materials are approximately electrically neutral, i.e., they contain nearly the same amount of positive and negative charges. Indeed, so powerful are electrical forces, that even when an object is electrically charged, for any degree of charge that any biologist will ever encounter, there is only ever a very small excess of one type of charge over another. As we shall see later on, the inability to accumulate significant amounts of charge of one kind without an approximately equal amount of charge of the other kind has fundamental implications for how electrical circuits operate.

In most conductors of electricity, there is one sign of charge that is mobile, and can move freely through the material, and the other sign of charge is fixed and locked into position in the material. In metals the mobile charge consists of electrons and the fixed charges are the positively charged atoms. In semiconductors there can be in essence two different sorts of materials, one where the mobile charge is electrons and the other where the mobile charge acts like a positive charge. The ions that can carry current in biological solutions can, of course, be of either polarity. Insulators are materials with no mobile charges.

For the most part, however, electronics does not require any knowledge of what the mobile charge carrier is. If negative charges go from A to B, or positive charges go from B to A, the net electrical effect is the same (although of course in electrochemistry this does matter).

Before we go any further, here are the common units seen in electronics:

Quantity	Symbol	Units	Abbreviation
Charge	Q, q	Coulomb	C
Energy	W	Joule	J
Voltage	V, v	Volt	V
Electric Field	E	Volts/Meter	V/m

Current	I, *i*	Ampere	A
Power	P	Watt	W
Frequency	f	Hertz	Hz
Angular frequency	ω	Radians/second	Rad/s
Resistance	R	Ohm	Ω
Conductance	G	Siemens ("Mho")	S, or Ω^{-1}
Capacitance	C	Farad	F
Inductance	L	Henry	H
Admittance	Y	Siemens ("Mho")	S, or Ω^{-1}
Impedance	Z	Ohm	Ω

The following prefixes and powers of ten are commonly used with these units:

Scaling Factor	Prefix	Symbol
10^{12}	tera	T
10^{9}	giga	G
10^{6}	mega	M
10^{3}	kilo	k
10^{-3}	milli	m
10^{-6}	micro	μ
10^{-9}	nano	n
10^{-12}	pico	p
10^{-15}	femto	f

Thus, 1000 volts is 1 kilovolt or 1 kV, and .001 amps is 1 milliamp or 1 ma.

Voltage

Voltage is always measured between two points. In some ways it is analogous to hydraulic pressure: the flow of a fluid in a pipe does not depend upon the absolute value of the pressure inside the pipe, but rather on the difference in pressure between the two ends. Similarly it is the difference in electrical potential between two points which is the "driving force" for the flow of electrical charge.

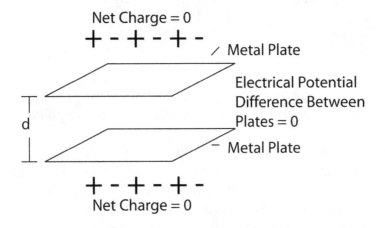

Net Charge = 0

+ - + - + -

Metal Plate

Electrical Potential
Difference Between
Plates = 0

Metal Plate

+ - + - + -

Net Charge = 0

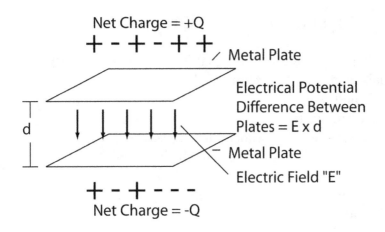

Net Charge = +Q

+ - + - + +

Metal Plate

Electrical Potential
Difference Between
Plates = E x d

Metal Plate

Electric Field "E"

+ - + - - -

Net Charge = -Q

Figure 2-1. Parallel conducting metal plates.

Consider two parallel metal plates separated by a good insulator like air or vacuum. If both plates are electrically neutral, the potential difference between the two plates is zero. However, if a specific amount of charge is moved from one plate to another, then there is a sort of tension in the system: connecting the two plates together with a wire would result in charges flowing through the conductive path which would bring everything back to neutrality. Positive charges would move to the negatively charged plate and negative charges would move to the positively charged plate.

In the condition of the charged plates, we say that there is an electrical field between them, and that the following relationship holds:

$$v = E \times d$$

volts = (volts/meter) x meters

An electrical potential difference exists between any two points in space.

A voltage has not only a magnitude but also a polarity. This means that, when specifying a voltage, we must be careful to define a reference polarity. Consider the generic network element in Figure 2-2. One terminal is labeled +, and the other is labeled -. This does not mean that the + terminal is more positive than the − terminal, rather, it means that we define the potential difference as:

v = (potential at + terminal) − (potential at − terminal).

So if v is positive the "+" terminal has a higher potential, and the "-" terminal has a lower potential, and vice-versa if v is negative. This care in defining reference polarity is needed both because we often cannot tell in advance which terminal will be positive and which will be negative, and because in many circuits the polarity of the voltage between two points can change over time. While we will not spend very much time on formal analysis of circuits, consistent labeling of reference directions is essential for any systematic analysis.

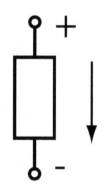

Figure 2-2.
Generic circuit element.

Current

Current is the flow of electrical charge. It is usually measured in units of amperes, or amps, which is the flow of one Coulomb of charge per second. There are about 6.24×10^{18} electrons/coulomb, so a single electron has a charge of about 1.6×10^{-19} Coulombs. Note from Figure 2-3 that the same polarity of current can be created by positive charges flowing in one direction or negative charges going in the other. From the point of view of electric circuit analysis, it doesn't matter. Electrons are negatively charged, but we

12

usually treat a current of electrons as carrying the same amount of positive charge in the opposite direction. Current is always defined as if it were the flow of positive charges, even though with metal wires it is negatively charged electrons that are flowing. Unless there is some other reason to worry about the nature of the charges themselves, as in electrochemistry, there is no reason to care about what species of charge is actually moving.

Voltage is always measured <u>across</u> a circuit element, and current is always measured <u>through</u> it.

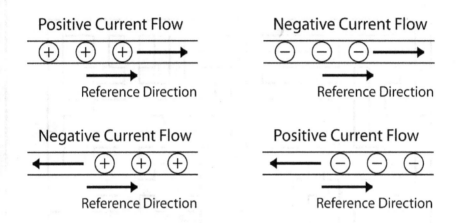

Figure 2-3. Current polarity and reference directions.

As with voltage, it is important to maintain consistent reference directions for current flow (see Figure 2-3). This figure illustrates arithmetically positive and negative currents using both positive and negative mobile charges. For a network element, the reference direction for current is always defined as being from the positive terminal to the negative terminal (see Figure 2-2).

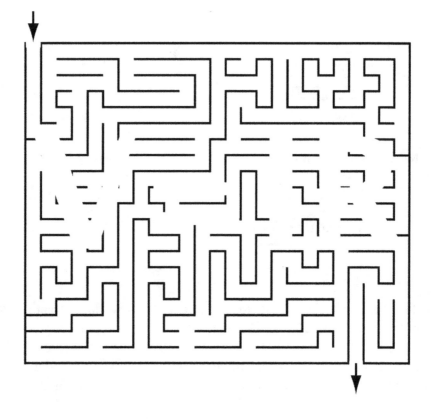

Electronics is confusing - or is it?
Remember the basic principles
(see chapter 5).

Chapter 3. Kirchoff's Laws

Figure 3-1 illustrates a basic four-element network (this being an electronics text, you could also call it a four-element circuit). For now we will not worry about exactly what the four elements A,B,C, and D are made of. For the purposes of the basic principles of Kirchoff's laws it doesn't matter. These elements could be batteries or resistors or light bulbs or hermit crabs, the same rules apply.

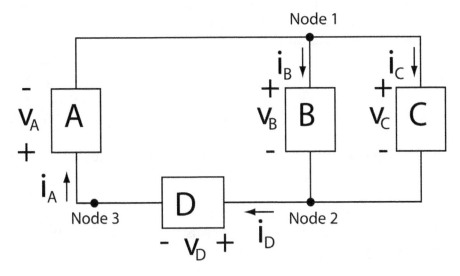

Figure 3-1. Generic Four-Element Network.

Note that this network has three *nodes*. Each node is a zero-dimensional point, the lines show only the patterns of connections and have no other meaning. For example, Node 1 connects the negative terminal of A to the positive terminals of B and C. That's it, it doesn't matter what order or direction we draw them in, it doesn't matter where on the lines connecting these elements we draw the little black dot, and indeed we could have outlined the entire set of lines connecting these parts and called that the node. Unlike real circuit elements made of real wires that are never perfectly conducting, we assume perfectly conducting wires and thus all points on each node have exactly the same electrical potential regardless of how long or twisty the lines are.

In real wires and elements the current will have some complex three-dimensional pattern as it flows though the wires and things of a circuit, but for the purposes of this circuit diagram we ignore all of that and only consider the net current flow through each point.

We say that elements B and C are connected *in parallel*. Because both ends are connected to the same nodes, they must therefore have exactly the same voltage across them (though the currents through them may be different). This is a key concept that we will beat to death with a stick before we are done.

Kirchoff's Voltage Law (KVL)

The algebraic sum of voltage drops taken around any complete closed loop in an electrical network is zero. Yes, <u>any</u> closed loop. The same circuit element can be part of several loops, and some loops can outline the perimeter of several smaller loops. So we could write:

$$v_A + v_B + v_D = 0$$

$$v_A + v_C + v_D = 0$$

$$v_B - v_C = 0 \qquad \text{(this can also be rewritten as } v_B = v_C, \text{ which is just elements in}$$
parallel)

Note that for this to work simply the loop has to go in the direction of + to – for each terminal. If the reference direction is opposite to the direction of the loop, you have to multiply that particular voltage by –1, as was done for the v_C term in the third equation above.

Kirchoff's Current Law (KCL)

As mentioned before, the attraction of opposite charges to each other is so powerful that there are normally only negligibly small imbalances between total positive and negative charges in any circuit element. This can be formalized in Kirchoff's current law, which states that the algebraic sum of all currents <u>entering</u> a node must equal zero. If the reference direction for a current is pointed away from the node, just multiply that current by -1. Hence, for the four-element network above we can write:

$$i_A - i_B - i_C = 0$$

$$i_B + i_C - i_D = 0$$

$$i_A - i_D = 0 \text{ (or more simply, } i_A = i_D).$$

The simplest case of KCL: when two circuit elements are in series with each other, it is always true that the current going through one element must equal the current going through the other one.

KCL is why they call it circuit theory: because electrical current always has to travel in loops or circuits, it can't be soaked up in any single place.

By writing these equations out, and if we also know the relationships between current and voltage, we can churn though N equations and N unknowns and solve for all currents and voltages explicitly. This is beyond the scope of this work, however, if you really need to tackle something that complicated, the techniques of doing so have become very advanced, and computer-based circuit analysis programs that do this all automatically are readily available as either commercial products or public-domain software.

Fortunately, you can do almost anything in electrical engineering without having to solve these equations yourself. Indeed, as we shall see later, you can do a lot by splitting a network up into elements in series and elements in parallel.

Ground

Voltage is always measured across an element. However, in complex circuits this can be tedious to specify. Thus, it is often the case that a single node will be selected as the common reference point for every voltage in the circuit. This reference point is called the *ground*, after the old days when it was literally a big copper bar pounded into the ground. It is often abbreviated as GND, or by a symbol that looks sort of like an upside-down Christmas tree. (Warning: sometimes the symbol that looks like an upside-down triangle is used to mean negative supply voltage instead, this varies).

Sometimes the ground really is a "true-earth ground", hooked up to the earth by something like a cold water pipe (hot water pipes go through hot water heaters and are therefore usually a less direct connection to earth

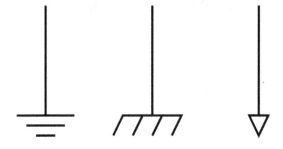

Figure 3-2. Common Ground Symbols.

ground). Often, however, the ground is simply the main reference point in a circuit, and with no connection to the earth the "ground" of a battery-powered circuit may have no relationship to true earth ground at all.

By convention, the ground is defined to be at zero potential. This is only for convenience: you could add any constant number to ALL the node potentials in a circuit and nothing would change, as it is only the voltage difference between two nodes that has any effect on current flow.

Using a ground can be confusing at first, but it does simplify the layout of a circuit.

Power

When a charge moves through an electric field, it either requires energy to move the charge against the field or yields energy when the charge moves with the field (as if a positive charge moves from the positive to the negative terminal). We define the power dissipated in a two-terminal circuit element as:

$P = v \times i$

watts = volts x amperes

It is important to maintain the convention that the direction of the reference current be from the positive terminal to the negative terminal. If P is positive, power is dissipated; if P is negative, power is delivered to the network.

Remember, power is in units of joules/second, or watts. To calculate the total energy, you have to multiply by time, for standard power systems we talk of kilowatt-hours.

Chapter 4. Sources

The first thing that we need to get an electronic circuit set up is something to get it moving, i.e., something to inject power into the circuit. Without a source of energy eventually every real circuit will run down to an equilibrium value where all currents and all voltages are zero. The two main "sources" used in electronics are voltage sources and current sources.

Voltage sources

Most commonly you will encounter voltage sources. An ideal voltage source is simply a network element with two terminals that maintains a specified voltage across its terminals regardless of how much current is flowing through it. Another way to look at it is that an ideal voltage source will supply whatever current is required to ensure that its voltage is maintained (we will cover what we mean by this in more depth later on). Common symbols for a voltage source are shown in Figure 4-1.

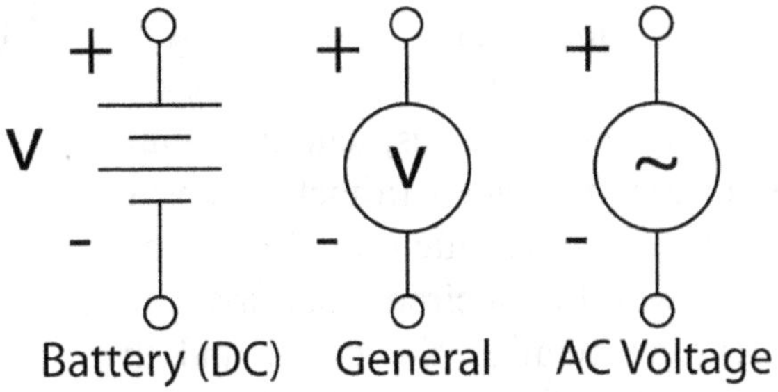

The symbol on the left is most typically used to indicate a battery, and more generally a source whose voltage is a constant function of time. The alternating parallel lines cartoon the parallel plates of different metals found in the first

Battery (DC)   General   AC Voltage

Figure 4-1 Voltage Source Symbols.

battery designs (the number of alternating plates does not usually have any relationship to the actual voltage – it's just a cartoon). It is important to realize that a voltage source may be a constant (or "DC", from so-called "Direct Current"), or it may vary as a function of time. The AC ("Alternating Current") in common use in houses is a voltage source whose voltage varies sinusoidally with a frequency of 60 Hz in the United States, 50 Hz in Europe.

It is slightly odd that "AC Voltage" is a generic term for sinusoisdally varying voltage, but then with AC current BOTH the current AND the voltage vary with time, so "AC" is often used as a generic term for time-varying. A voltage source may also vary as a more complex function of time, for example, the output of a compact disk audio player can be treated as a voltage source whose voltage varies over time as a function of the audio waveform.

All real-world voltage sources depart from the ideal in one respect or another. For example, while an ideal voltage source could deliver any amount of current, real sources are limited to some level of power they can deliver before they "run out of oomph" and deliver less voltage than desired. Also, real voltage sources like batteries decay over time with use. As we shall see in later sections, these deviations from the ideal can be modeled by adding additional electrical circuit components to an ideal source.

Current Sources

The other source is a current source. A current source maintains a specified value of current (either constant or a function of time) regardless of the voltage across its terminals. Common symbols for a current source are shown in Figure 4-2.

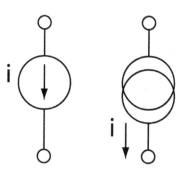

Figure 4-2. Current Source Symbols.

Unlike voltage sources, current sources are not easy to make. In fact, a constant-current source is often implemented in the real world by a circuit that first senses the current running though it, and then dynamically adjusts the voltage across its terminals so as to cause the desired amount of current to flow.

As such, current sources are not usually part of models of natural phenomena, but instead are most often found as complex integrated circuit devices built for their value in certain kinds of signal processing, in driving different electrical devices, or in electrochemical systems where you want to deliver a set amount of charge per unit time even though the resistance of the circuit is changing ("current clamp").

5. Linear Resistive Networks

The most basic circuit element is the *linear resistor*. The resistor part means that the circuit element is not a perfect conductor but offers some degree of resistance to current flow; the linear part means that (in this simple case) the current-voltage relationship is a straight line (see Figure 5-1).

The symbol of resistance is R. The relationship between current and voltage in an ideal linear resistor is just

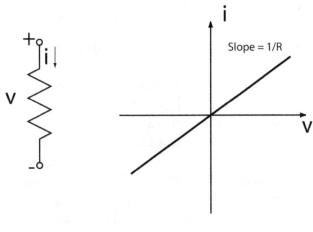

Fig. 5-1. Ideal Linear Resistor.

$$V = IR$$

This equation is known as <u>Ohm's Law</u>. This is not really a law, as there are other circuit elements that do not have this relationship between current and voltage, so this is really an operational definition of resistance. But most people are not this picky and just refer to it as Ohm's law. It can also be written as

$$I = GV$$

Where G is the conductance, which is the reciprocal of resistance

$$G = 1/R$$

R is given in units of ohms, abbreviated Ω, and conductance is given in units of Siemens (or archaically, "mhos", or Ω^{-1}).

Power Dissipation in Ideal Resistors

As defined previously, the power dissipated by a two-terminal circuit element is the product of the current and the voltage. This holds for all

21

elements, including resistors. However, because current and voltage have a simple linear relationship in an ideal resistor, we can do a little algebra and get:

$$\text{Power}_{\text{resistor}} = i^2R = v^2/R$$

Note that <u>this</u> relationship only holds for resistors.

The fact that the power is related to the square of either the voltage or the current has important practical implications. In particular, it makes computing the average power delivered to a load via an alternating current somewhat difficult. You cannot use the average of a time-varying voltage or current to compute average power. Instead, you square the voltage (or current), take the average, and then take the square root. This is the Root Mean Square (RMS) value, and you can use the RMS value exactly as if it were a constant (DC) value for the purposes of computing power. (For a sinewave, the current and the voltage reverse polarity together, so the instantaneous power is always positive).

For a DC voltage the RMS value is just the DC voltage unchanged. For a sinusoidal alternating current, the RMS value is 1/sqr(2) times the peak voltage. Alternatively, the peak voltage is sqr(2) times the RMS voltage. Standard American house current is 115 volts AC RMS, so the true waveform of voltage as a function of time is a sinusoid with a frequency of 60 Hz and a voltage that oscillates between −163 and +163 volts peak-to-peak. For more complex waveforms (such as the audio waveform delivered to a loudspeaker) there are no simple formulas and you need a meter than can calculate a <u>true RMS</u> value of voltage or current.

Real-World Resistors

There are two basic kinds of real world resistors. There are resistors designed specifically for use in circuits to deliver a precise amount of resistance. These devices are called <u>resistors</u> (obviously), and they are so close to ideal resistors that for the most part people don't bother with the "ideal" or "actual" qualifiers. For the perfectionist there are many subtle differences between real and ideal resistors, but the primary deviations from the ideal occur at either very low voltages and currents, where quantum effects can become important, or at very high voltages and currents, where power dissipation in the resistor can heat it up which in turn changes the resistance, and either cause it to melt or, in the extreme case, explode.

Thus, other than the value of the resistance itself, real resistors are usually specified by their ability to dissipate heat. In a typical purpose-built circuit-element resistor all of the power delivered to the resistor by the rest of the circuit is dissipated as heat. Resistors designed for signal processing and computer boards most often have power ratings of 1/8, 1/4, or 1/2 of a watt: larger resistors are used in various higher-power systems and high-power rated resistors can be used as dummy loads in power generating systems, these latter may have elaborate radiators to help dissipate the heat and they may be very large indeed.

The other kind of real-world resistor consists of various elements such as light bulbs, loudspeakers, electric motors, etc., and they are often modeled as resistors simply because it is so easy to do so and for most practical purposes the modeling works pretty well. For example, even though a personal computer contains many complex circuits, as far as the power system that it is plugged into is concerned it can be modeled as a simple resistance that draws a specified current at the rated voltage, and dissipates the current times the voltage watts of power.

It is obvious but somehow counterintuitive that low-power devices are modeled as large resistances, but high-power devices are modeled as small resistances.

Single Resistor-Source Network

Figure 5-2 illustrates a single-resistor network, where a battery is hooked up in parallel to a resistor with resistance R. The voltage created by the voltage source, V_B, is exactly the same as that seen across the resistor, V_R. The current flowing through the

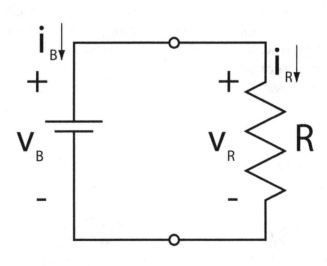

Figure 5-2. Single-Resistor Circuit.

resistor is $i_R = v_B/R$. This is also exactly the same in magnitude as the current flowing though the voltage source, i_B. Note that, by convention, the reference direction for the current flowing though the voltage source is drawn going from positive to negative. Thus, as current is actually flowing from – to +, the magnitude of v_B is actually the opposite in sign from that of i_R. Among other things, this makes calculations of power dissipation come out right. The resistor dissipates $v_R \times i_R$ watts of power, but the voltage source dissipates $-v_R \times i_R$ watts, which is another way of saying that the voltage source supplies $+v_R \times i_R$ watts to the circuit.

For this circuit, as with most circuits you will encounter, there is no need to be so fussy with reference directions, as it will be obvious what elements are supplying power and what elements are absorbing power. Still, there are times when intuition can be misleading and it never hurts to know the mathematically precise way of analyzing a circuit.

Note also, if v_B should become negative, either because the battery was reversed or the input voltage polarity varied as a function of time, that everything would also work out. The direction of current in the network would reverse, but the power calculations would remain unchanged because a negative current times a negative voltage is still a positive power.

Key point: ideal resistors have no memory. Another way to say the same thing is that ideal resistors do not store any energy. If you change the voltage applied across an ideal resistor, the current changes exactly in step in accordance with Ohm's law, with no delays or oscillations or other funny stuff. If the input voltage is a complex function of time, for example a voltage waveform representing an audio speech recording, the current as a function of time will be exactly the same, subject only to the scaling of Ohm's law.

Resistors in Series

One of the two circuits which are the bedrock on which much of electrical engineering is ultimately based is *resistors in series*.

In Figure 5-3, two resistors, R_1 and R_2, are placed *in series* with the voltage source. From Kirchoff's Current Law we know that $i_1 = i_2 = -i_B$. From Kirchoff's Voltage Law we know that $v_1 + v_2 - v_B = 0$. However, we typically write the voltage source on one side of the equation and the

24

resistors on the other, or $v_1 + v_2 = v_B$. Informally, the voltage of v_B is split up across R_1 and R_2. If you do the math, you find out that the combination of two resistors in series acts exactly like a single big resistor whose value is just the sum of the two separate resistances. In effect, you analyze this circuit just like you would the single resistor circuit seen previously, but setting the value of $R = R_1 + R_2$. Thus, the current though the network is just $I = V/R = v_B / (R_1 + R_2)$. Now that we know what the current though the entire circuit is, we can go back and figure out what v_1 and v_2 are by using Ohm's law, so $v_1 = I \times R_1$, and $v_2 = I \times R_2$.

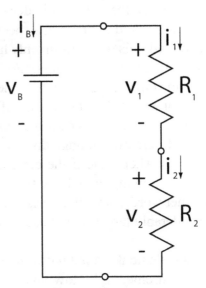

Figure 5-3. Two Resistors in Series the Formal Textbook Way.

Most of the time you will not see circuit diagrams drawn like this. Instead, you will see them as drawn in Figure 5-4. It is understood by inspection that all the current goes through all elements, and that the battery delivers power and the resistors dissipate it, so there is no need to label all the reference directions.

Also, in most cases we will talk about the voltage at a node (in this case v_1 and v_2): it is understood that these voltages are really all with respect to ground, and it is also understood that all grounds are connected together. Drawing this circuit in this way makes it easier to write the equation for voltage v_2:

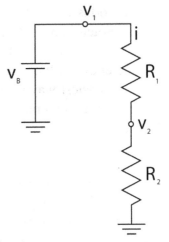

Figure 5-4. Two Resistors in Series Drawn the Usual (Easy) Way.

$$v_2 = v_1 \times R_2/(R_1 + R_2)$$

This is called the *voltage divider* relationship. The voltage at v_2 is the voltage at v_1 divided by the relative sizes of the resistors. If $R_1 = R_2$, then v_2 is just 50% of v_1. If R_2 is three times R_1, then v_2 is 75% of v_1.

25

You will see the voltage divider used a LOT in circuits. It is a simple and reliable method of creating a specific voltage from a higher one. However, be warned: this relationship only holds if no current is drawn out of the node connecting the two resistors. For example, if in this figure we had hooked up some other element to the node at v_2, the current drawn off through that element would alter the balance of currents in the circuit (the current though R_1 would no longer equal the current though R_2) and change the voltage at v_2. In practice voltage dividers can only be used to create a voltage for use somewhere else if the current drawn off is considerably less than the currents flowing though the two resistors that make up the voltage divider. In other words, any other resistances hooked up to a voltage divider must be considerably larger than those in the divider circuit.

Circuit elements called *potentiometers* take the concept of a voltage divider to an extreme, and allow you to "pick off" any voltage ranging from the supply to ground (see Figure 5-5). These are built as resistors that are either circular or straight (linear), and that have a moveable conductive wiper that slides along the resistive element. Digital controls are starting to become more common, but the next time you twiddle a knob on a stereo odds are still that you are using a potentiometer to change a parameter of a circuit. Using a potentiometer as a voltage divider is probably the easiest way of using a manual control to set a parameter in an electric circuit.

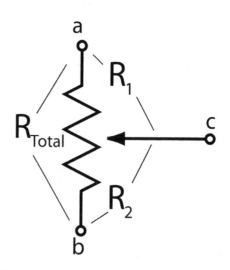

Figure 5-5. Potentiometer.

Note that the total resistance between the two end terminals **a** and **b** never varies, and that as the wiper is moved it is always true that $R_1 + R_2 = R_{total}$. If you only hook up the "wiper" terminal and one other then you have a simple variable resistor whose value changes from 0 to R_{total}.

Resistors in Parallel

The second classic circuit is resistors in parallel. If resistors in series can be

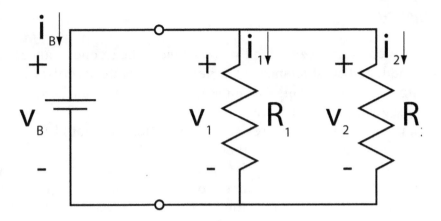

Figure 5-6 Resistors in Parallel.

considered to be a voltage divider, then resistors in parallel can be considered to be a current divider. As with a river that forks into two streams, Kirchoff's current law states that all the current flowing into two resistors must be divided between them with nothing left over. As with a river, the largest current flows through the path with the lowest resistance.

Note that this circuit has only two nodes. Thus, the voltage across all elements is the same, $v_B = v_1 = v_2$.

We can easily solve for the currents in the resistors via ohm's law:

$i_1 = v_B/R1$

$i_2 = v_B/R2$

Using Kirchoff's current law we then know that $-i_B = i_1 + i_2$ (We are being fussily academic about the sign here: in a case as simple as this the direction of current flow is so obvious that saying that the total current coming out of the battery is the sum of the current in R_1 and the current in R_2 suffices).
If you do the math, you can prove that two resistors in parallel act like a single smaller resistor whose value is given by:

$R_{total} = R_1 \| R_2 = R_1R_2/(R_1 + R_2)$

Where the notation "$R_1 \| R_2$" means "R_1 in parallel with R_2." If $R_1 = R_2$, then the total resistance of both in parallel is $1/2\ R_1$. This makes sense, as there is twice as much "freeway" for the electrons to travel on.

Conductances

If you use conductance instead of resistance, then conductances in parallel add, and conductances in series use the same sort of formula as resistors in parallel. This can sometimes make analyzing a network easier, especially if you have more than two resistors in parallel. For example, if you had 50 resistors in parallel, you could find the equivalent resistance of two of them, then combine this resistance in parallel with that of another resistance, and repeat 48 times. Or you could just add the 50 conductances together and get the equivalent conductance of all 50 resistors together – take the reciprocal of this equivalent conductance to get the overall equivalent resistance.

Typically engineering diagrams use resistance, but electrical models of biological systems use conductances, because it simplifies the analysis of biologically important situations such as many ion channels in parallel across a membrane.

More complicated topologies: If circuits get really twisted or complex there is often no alternative to using KVL and KCL, writing out all the simultaneous equations and solving – or even better, getting a hold of a circuit-simulating program and having it do the work for you.

However, there are many problems one step up in complexity from two resistors in series or two resistors in parallel that it is useful to be able to solve yourself, if only for the increased intuitive understanding of electric circuits. You just combine elements alternately in series and in parallel until you get the overall equivalent resistance, solve for the voltage and current across the entire network of resistors, and then proceed back down the chain solving for the currents and voltages in the individual elements as you go. A specific example of this is shown in Figure 5-7.

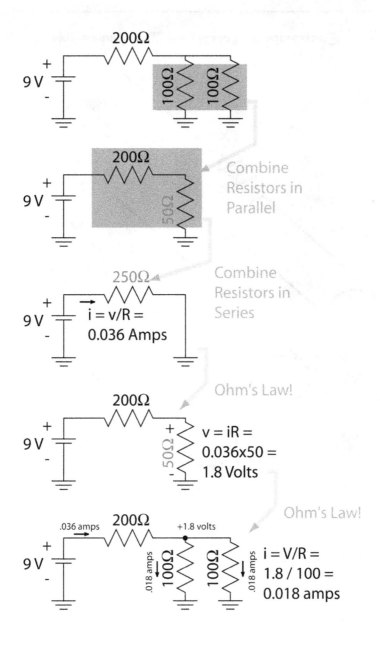

Figure 5-7. Solving a circuit with multiple substitutions.

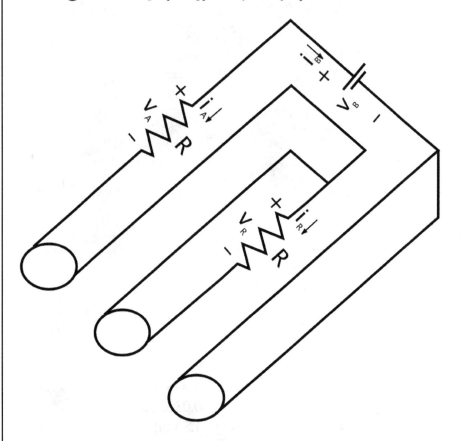

Sometimes the topology of an electrical circuit can be confusing as drawn. You can always re-draw it, as long as you keep the same pattern of connections between elements

6. Linearity, Superposition, and Equivalence

We have already made reference to the fact that ideal resistors are *linear* devices. The plot of current vs. voltage for such a device is a straight line, but this is not the true definition of linearity. For a function to be linear, it must satisfy the constraint of superposition, specifically,

$$f(x_1 + x_2) = f(x_1) + f(x_2)$$

Another property of linear functions is scaling, where "a" is a constant:

$$f(ax) = af(x)$$

So twice the input gets twice the response, -4 times the input gets –4 times the response etc.

The entire issue of linear systems and their analysis encompasses a vast amount of the field of electronics and signal processing. When a system is linear there are many powerful techniques that can be applied to analyze it. Many natural processes are either nearly linear, or can be approximated as being linear over some restricted range of inputs and outputs. Electronic systems are often designed to be as linear as possible to facilitate their use. For example, transistors are inherently nonlinear devices, but considerable effort has been expended to find ways of using transistors to build amplifiers and filters that behave in a highly linear fashion.

Assuming linearity greatly simplifies the analysis of electrical circuits. This is not a priority for a biologist; however, understanding the implications of linear behavior is critical in dealing with many important biological systems, such as the spread of electrical activity in axons and dendrites. It also crops up in many areas not related to electronics that are nevertheless of great importance to the study of biological systems – for example one can consider simple cells in primary visual cortex to be linear spatial filters of an image.

In this section we will deal exclusively with the circuit-theoretical aspects of linearity, and we will cover the more general implications in a later section.

Superposition

Suppose one needs to analyze a complex circuit with many resistors and with many different sources (voltage or current) connected up in a complex manner. It could seem daunting. However, if all the circuit elements are linear, it turns out that we can solve the network for each source separately, and the final solution is just the sum of the solutions for each individual source.

In other words: to analyze a network with two batteries, first analyze the network with one battery in place and the other one turned off, then the opposite, and add the results.

To turn a voltage source "off", replace it with a short circuit (wire), as this forces the voltage to be zero. To turn a current source "off", replace it with an open circuit (no wire), as this forces the current to be zero. Figure 6-1 shows an example of this process. You can do this with any number and any mixture of current and voltage sources. Just remember, if there is even one non-linear element in the circuit this technique is not guaranteed to work.

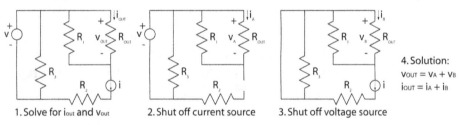

Figure 6-1. Using superposition to solve for multiple sources.

Equivalent Circuits

We have already seen that we can often simplify a network by taking resistors in series or in parallel and treating the two together as if they were a single resistor with a value that is a function of the two separate resistors. We can extend this concept to more elaborate combinations of elements.

First, for any network of ideal linear resistors, no matter how complicated, for connections at just two points you can always treat the entire circuit as just a single equivalent resistance. Calculating the value of that resistance might be hard, but you can always do it – and for a real network of resistors you can always measure it with a resistance meter.

32

When voltage and current sources are involved, things get a little more complicated, but not much. It can be shown that, for any combination of resistors, current sources, and voltage sources, you can always replace them with either one voltage source and one resistor in series, or one current source and one resistor in parallel (see Figure 6-2). For the Norton Equivalent circuit i_{SC} is the current that flows out of the circuit when it is short-circuited (the terminals connected by a wire), for the Thevenin equivalent circuit v_{OC} is the voltage across the terminals when the circuit is not connected to anything (open circuit).

There are a lot of theorems that engineers like to derive in regard to these equivalent circuits, but you don't need to know any of them. What is important is that many complex pieces of instrumentation, and many interesting biological processes, can be modeled

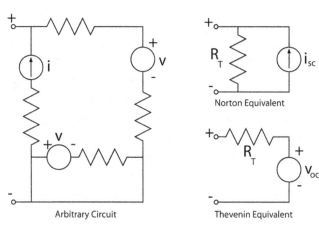

Figure 6-2. Equivalent Circuit Concepts.

by these circuits. Even when the modeling is not exact, it is often a good approximation. Also, you need to know when you see a bunch of resistors and batteries wired up by themselves in a complicated way that, *as seen from two connecting terminals,* the circuit can not have any behavior more complicated than a single battery and resistor in series. Knowing this can keep you from getting overwhelmed by things that are in truth simple.

The concept of an equivalent circuit is a critical one for biological applications. It often lets you model a complex cell, with many pieces of membrane and many thousands of separate ion channels and pores, with just one battery, one resistor, and one capacitor (we will cover capacitors later). Also, complicated pieces of electronic measuring equipment can be modeled as just a single resistor at the input to the equipment, this is an absolutely vital concept for understanding how to make accurate measurements of weak signals.

Critical point: the equivalent circuit is only valid when "looking at" the circuit from the specified two connections. "Look at" a circuit from another two connections and you will get a different equivalent circuit. Additionally, even though two circuits may be electrically indistinguishable (i.e., *equivalent*) as seen from two specific terminals, the internal flow of current, total internal power dissipation, and operation of different pieces of the circuits may be quite different.

As one example of the use of equivalent circuits, real physical batteries are not perfect voltage sources. Accurate physical models of batteries are very complicated, and you could fill volumes going over all the subtle points. However, for many purposes it suffices to model a real battery as an ideal battery in series with a single resistor – the Thevenin equivalent circuit. Like a real battery, this equivalent circuit delivers maximal voltage to an open circuit, and its voltage declines as it is forced to deliver more current. This roughly captures the limited ability of any real battery to supply power. In the case of batteries we do not always talk of its "Thevenin equivalent resistance", but more commonly, of its *internal resistance*. Batteries that can only deliver small amounts of current are modeled as having large internal resistances, and batteries that can deliver large amounts of currents can be modeled as having small internal resistances.

This concept is also a key one for any instrumentation application. If you have an amplifier that is designed to accept a voltage signal from somewhere, if it is to accurately represent what it is amplifying, the "input resistance" of the amplifier must be considerably larger than the "output resistance" of the circuit/device/biological process that is supplying its input. Additionally, the "output resistance" of the amplifier must be low relative to the input resistance of whatever its input is connected up to.

In general: amplifiers and other general-purpose pieces of electronic instrumentation will have very high input resistances, to minimize the effect on what is being measured, and as low an output resistance as possible, so that the output voltage is least affected by what it is hooked up to.

In many cases where small voltages are amplified there is frequently a small preamplifier used before the main amplifier. Often this preamplifier has very low amplification – in fact it may have no amplification at all. So why bother with a preamplifier that outputs the same voltage that it has at its input? The answer is that the preamplifier will have a very high input resistance and low output resistance. An amplifier with no voltage gain is often called a voltage follower (because the voltage at the output *follows* the output at the input), or a *buffer*. Preamplifiers for small signals are sometimes also called "headstages", they may have either unity or small (10x) voltage gains, but they always have high equivalent input resistances and low output resistances. By buffering the small input signals as close to the source as possible, a preamplifier can send signals down long cables with less noise than the raw unbuffered bioelectrical signal, with its high equivalent resistance, can.

Biological systems such as single neurons can be modeled as voltage sources with very high internal resistances (you can't power a 60-watt light bulb with a microelectrode hooked up to a single nerve cell, now can you?) – therefore, any amplifier that is to accurately measure the signals coming from a single neuron (or any cell) must have a very high input resistance (like 10^{12} Ohms).

In the example in Fig. 6-3, we have a typical recording setup where a microelectrode is inserted into a cell, the signal is amplified, and then recorded via a computerized data acquisition system. A simplified circuit model is at bottom.

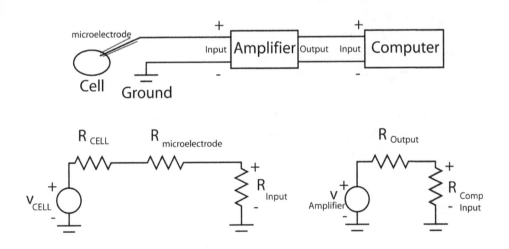

Figure 6-3. Biological instrumentation (top) and equivalent circuit model (bottom). The amplifier and computer are very complex devices, but from the perspective of their inputs and outputs they can be modeled as large resistors. You can then use the voltage divider relationship to determine how the amplifier characteristics may affect your measurements. This is a critical concept! *Go no further until you understand it!*

35

The electrical properties of the cell can be *modeled* as a voltage source in series with a (very large!) resistance. The microelectrode can be *modeled* as another (large!) resistor. The input of the amplifier is *modeled* as a resistance, and the voltage output $v_{Amplifier}$ of the amplifier is A_0 x the voltage across R_{Input}, where A_0 is the gain (amplification) of the amplifier. R_{Output} is the internal resistance of the amplifier output, and $R_{Computer\ Input}$ is the equivalent input resistance of the computerized data acquisition system: the computer records the voltage across this resistance. From our old friend the voltage divider, it should be apparent that we want $R_{Input} \gg R_{CELL} + R_{microelectrode}$ and $R_{Computer\ Input} \gg R_{Output}$. If this is not the case then there will be little voltage across the measuring parts of the circuit, and the recorded signals will be very small – perhaps so small that they are totally obscured by noise.

Note that the inputs of amplifiers are not really big resistors, but they can be *modeled* as resistors where the amplifier responds to the voltage across it.

Warning: many inferior texts claim that the entire voltage drop occurs across the resistance of the microelectrode. Not necessarily true! The cell (or membrane patch) itself is limited in its ability to deliver current to an external circuit. R_{CELL} may or may not be larger or smaller than $R_{microelectrode}$. Thus, the microelectrode might or might not be the limiting factor in your recording setup, all you can really say for sure is that the equivalent resistance of the cell and microelectrode is at least equal to that of the microelectrode. In practice it is difficult to get specific numbers for the equivalent resistance of a cell as seen from a microelectrode, so it's safest to use amplifiers with really, really large input impedances, more than would be required by just the impedance of the microelectrode per se.

It is typical to use voltage to transfer signals in circuits, almost all garden-variety amplifiers, oscilloscopes, filters, data acquisition systems etc. are set up in this way. Nevertheless, using voltage to transmit a signal is not obligate, and there are systems that use current to transfer signals – some industrial and serial line communication systems do this. However, in this text, unless otherwise indicated, we will assume that voltage carries the signal.

Impedance

There is a generalized form of resistance called *impedance*. It has the units of ohms, just like a resistor; however, it has additional complexities that let it handle time-varying signals. This is far too involved a subject to cover in this text. For now, just know that: for circuits with no energy-storage elements, impedance is precisely resistance – and even when it's not, just think "sort of like resistance, but that takes timing into account/depends on frequency" and you won't be far wrong.

Thus: it is typical to speak of *high-input impedance* amplifiers. When you hear this word, think *high equivalent input resistance* and you'll have it about right.

Admittance

The reciprocal of impedance is admittance, which is measured in units of Siemens. Admittance is to conductance what impedance is to resistance.

Why does that coaxial cable say "75Ω" on it?

Often in labs you will use co-axial cables to hook up instruments. This is a cable where one conductor is just a wire, and the other is a hollow cylinder separated from the inner wire by an insulator. The outer conductor is usually the ground. As we shall see later, this arrangement tends to reduce electrical interference.

These cables are typically stamped with a resistance value such as 75Ω (or 50Ω, or 300Ω, or whatever). However, if you use an ohmmeter, you will find that the resistance of either wire is very low, and the resistance between the two wires is very high. Where is my 75Ω?

The answer is that the 75Ω is the impedance of the cable to a signal that changes very fast. Before the electrical signal at one end can make it to the other end, the interactions of the electrical and magnetic fields inside the cable makes it look to an external circuit as if it were just a 75Ω resistor. It is only after the signals have reflected back and forth down the cable that the resistance between the two conductors appears to be high.

For almost all purposes of biological instrumentation: ignore the 75Ω, treat the cable like two separate low-resistance wires. However, be warned that, for low-frequency applications, coaxial cable acts like it has a capacitor connecting between the two conductors that has a value of (variously) around 30 pF (pico-Farads) per foot of cable. This *stray capacitance* means that it is a very bad idea to hook up a microelectrode or other high-impedance source to a length of coaxial cable, because the signal can be absorbed by the capacitance (more on this later).

Addendum: it turns out that if the equivalent resistance of the circuit that the cable is hooked up to matches the equivalent resistance of the cable itself ("impedance matching"), then there are no echoes back-and-forth down the cable, something that is important for high-speed circuit operation, such as in radio, television, and computer backplane operation. A cable that has a matching equivalent resistor at its end is said to be *terminated*.

Measuring Stuff

There are three basic measurements that are made with a standard hand-held meter: voltage, current, and resistance. The ideal voltmeter is modeled as an open circuit (infinite resistance), real voltmeters have some large but still finite resistance and the voltage "dropped" (measured) across the resistance is where the voltage reading is taken from. As before, voltage is always measured between two points. The larger the effective resistance of the voltmeter, the less the presence of the voltmeter disturbs the circuit being measured. Most hand-held meters do not have sufficiently high input resistance to directly measure the membrane potential of a single cell.

The ideal current meter is modeled as a wire (zero resistance). Real current meters always have some small but still finite resistance. Obviously, the smaller the resistance of a current meter ("ammeter"), the less effect the current meter will have on the circuit under test. Current is always measured through an element, so to measure current you have to unhook the element under test and put the ammeter in series with it.

Those multi-purpose meters can be easily switched between reading voltage and current but beware! Use a meter set to measure current as a voltmeter and you will create a short-circuit, and likely blow a fuse or maybe damage something.

Resistance is a trickier thing to measure, all resistance meters operate by applying a set voltage (or current) and measuring the resultant current (or voltage). Thus, when a meter is set to measure resistance it is going to actively inject current into whatever it is hooked up to. Depending upon the meter, this could damage the tissue/device under study, and at the very least it will perturb the operation of any functioning circuit hooked up to the device under test. Thus, you cannot use a simple hand-held meter set to measure resistance to determine the equivalent input or output resistance of a complex device, or even to determine the value of a resistance while it is embedded in an operating circuit! Best use the resistance meter to check the resistance of isolated resistors or wires that are removed from a functioning circuit.

7. Capacitors

Capacitors are second only to resistors in their relevance to biologists. Capacitors are very commonly found in electronic instrumentation, and many biological processes can be modeled as capacitors. The cell membrane is a prime example of something that can be profitably treated as being a capacitor. Additionally, many forms of noise or interference can be modeled as an unwanted "stray" capacitance.

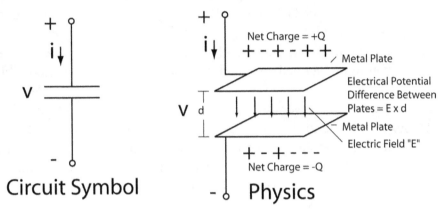

Figure 7-1. Capacitor basics.

The symbol of a capacitor is two parallel lines (see Fig.7-1), which is a cartoon of the simplest capacitor geometry of two parallel plates (real capacitors can have many different geometries, but the overall circuit properties are the same).

A capacitor stores electrical charge on its plates. The amount of charge stored is just:

$$Q = Cv$$

Where Q is the amount of charge stored, C is the capacitance (typically in Farads, abbreviated "F"), and v is the voltage across the capacitor. When a capacitor stores a given amount of charge, there is actually a balance of positive and negative charges on the opposing plates, so the overall charge on the entire capacitor is zero.

There are 1.6×10^{-19} coulombs/unit charge (or per *univalent* ion), or one Coulomb consists of about 6.24×10^{18} charge units (protons or electrons). Typical cell membranes have a capacitance of about 1 $\mu F/cm^2$, or 0.01 pF/μ^2.

39

Capacitors, like ideal resistors, are linear devices (more on that later!). However, unlike resistors, capacitors can store energy, that is, they have a memory. Thus, you cannot simply draw the current-voltage plot in a two dimensional plane. Instead, the relationship between current and voltage depends on time, specifically,

$$I = C\, dv/dt$$

In other words, the current flowing into or out of a capacitor is a function of the rate of change of the voltage. If the voltage is constant, there is no current flowing though a capacitor. This makes sense when you consider that as a capacitor is basically two conducting plates separated by an insulator, therefore sustained current flow through a capacitor is impossible.

A capacitor can have power delivered to it by an external circuit, in which case it stores the energy in the electric field between the two conductive plates. A capacitor can also use up the energy stored in it to deliver power to an external circuit. An ideal capacitor is a lossless device that can transfer energy back and forth without dissipating any of this power inside it as heat.

Because of the dependence upon time, circuits with lots of capacitors (and other energy-storing elements) can be much harder to analyze than those that only have memoryless resistors. Nevertheless, there are some simplifying properties.

You can replace capacitors in parallel with one larger capacitor whose capacitance is the sum of the separate capacitances. Think of sticking a bunch of parallel-plate capacitors together side-by-side to get one larger capacitor. You can also replace two capacitors in series with one smaller capacitor whose value is given by

$$C_{series} = C_1 C_2 / (C_1 + C_2)$$

(Capacitors in series use the resistors in parallel formula).

Real-World Capacitors

As with resistors, real capacitors come in two basic varieties: the kind that is purpose-built to serve as an electronic circuit element, and the kind that appears as part of a natural process such as a biological membrane. Real-world capacitors are specified primarily by their capacitance measured in Farads, although a Farad is a lot of capacitance so it is more common to see micro-Farads μF or pico-Farads pF ("puff") in electronic devices. Purpose-built capacitors are also specified by their voltage rating, which is the voltage at which the insulator between the conductive plates "breaks down" and the capacitor fails. As you can store more energy in a capacitor of a given value if you increase the voltage, there is a tradeoff between capacitance, voltage rating, and physical size. A capacitor that can store a Farad at 1000 volts is going to be LARGE, but a capacitor that stores a Farad at 1 volt can be 1/1000 the size, and a capacitor that stores 1 pF at 200 volts can be fairly small.

Purpose-built capacitors vary in many parameters, and electronics texts list all the different specialist types. All we really need to know is that circuit-type capacitors come in two basic flavors: unpolarized, that behave quite close to an ideal capacitor and that are used for signal processing and precision instrumentation, and polarized, or electrolytic capacitors, that are big ugly brutes that store a lot of energy in a compact space but that have really poor signal-processing properties (add distortion to a signal), they have trouble responding to rapidly-varying signals and they limit a design by requiring that one terminal always be positive relative to the other. Electrolytics are thus most commonly used for tasks such as smoothing out the ripples in a power supply or storing small amounts of energy for things like electronic flashes in cameras.

The analysis of time-varying signals and circuits that involve capacitors can be very complicated, but there are two simple heuristics (rules-of-thumb) that can be used.

=> When considering signals that are either constant ("DC") or very slowly varying, remove all the capacitors from the circuit (that is, replace all the capacitors with open-circuits).

=> When considering signals that change very rapidly (how rapidly is very rapidly depends on the circuit properties), replace all capacitors with short-circuits e.g. wires.

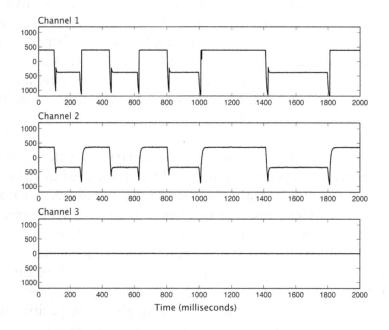

Shown are three channels worth of data taken at the same time from an Analog-to-Digital converter. The data were sampled at a rate of 1000 samples/second (1kHz) for two seconds. Channel 1 was hooked up to a real source of data. However, channel 2 was left unconnected to anything, and channel 3 was grounded, i.e., the signal input was shorted out to the ground. Note that the grounded channel correctly reads zero, however, the unconnected channel looks like a distorted copy of the channel with the real data on it. This is due to capacitive coupling between the channels.

If you accidently used the data from channel 2 in your analysis you could have something that looks like data but it will not be accurate. With more complex waveforms it can be even harder to tell. Ground unused inputs, and check which wire goes where! Remember also that some systems use channel 1 as the first channel, and some use channel 0 (zero), further increasing the possibility of confusion. Always run test patterns and check!

8. Single Time-Constant Circuits

Congratulations! You are almost half-way to becoming an electrical engineer. Well not really, but if you can get past this chapter you will have a lot of the fundamentals that a biologist needs. It's really amazing what you can do with just the techniques listed up through this chapter.

In figure 8-1, we consider a capacitor that is initially completely discharged. Remember, a capacitor has a memory, so we need to specify its initial state via the single *state-variable* of its charge. As a state-variable, the charge encapsulates all of the capacitors' past history, nothing else need be known.

Now at time t=0 we change the input voltage from 0 to "A" volts. You can't change the voltage across a capacitor until you have pumped current into it for a while, so the voltage across the capacitor stays low. Thus, all the voltage after the sudden change is across the resistor, and the current flowing into the capacitor is the same as the current flowing though the resistor = A/R. However, as the capacitor starts to charge up, its voltage rises. Thus, there is less of a voltage drop across the resistor, and less current going through it, which reduces the rate at which the capacitor acquires charge, which reduces the rate at which the capacitors' voltage approaches A.

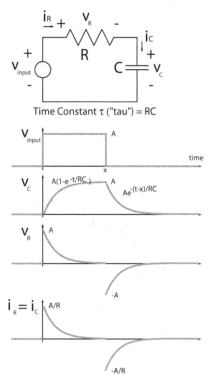

Figure 8-1. Single Time-Constant RC Circuit.

When the input voltage is reduced to zero, the direction of current flow reverses as the capacitor discharges though the resistor.

The simple way to do it: immediately after a sudden change a capacitor acts like a wire, after things have settled down a capacitor acts like an open circuit. With a single capacitor in the circuit you connect the start and the end with an exponential rise or fall and you are done.

This process is a classic example of exponential decay/rise – it's sort of like half-life and radioactive decay. However, the math works out better if instead of talking about half-life we talk about *time constants* (see figure 8-2). The time constant τ ("tau") for any circuit with a single resistor and a single capacitor is just RC, the response to a sudden change in voltage will be some variation on the theme of $e^{-t/\tau}$.

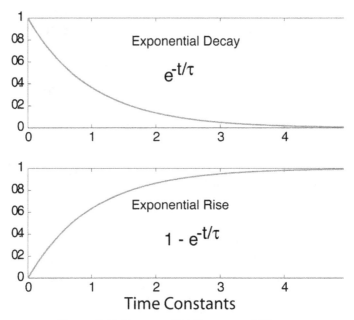

Figure 8-2. Exponential Decay and Rise.

The number "e" is an irrational number like π, and is approximately 2.7183. It crops up all the time in engineering, because the derivative (rate of change) of e^x is itself e^x.

In the initial part we know that the voltage across the capacitor will start at zero, it will then rapidly rise towards voltage A but the rate of rise will slow and it never actually gets all the way to A. Thus, the voltage across the capacitor will be

$v(t) = A \ (1 - e^{-t/RC})$

At every time constant's worth of time, we have gone to within $1/e \approx 0.37$ of the final value from where we started. At five time constants we have traveled to within $(0.37)^5$ of the final value from where we started, which is to within 0.7% of the final value. It may officially never get there, but after a few time constants it gets pretty darned close.

9. Inductors

Inductors are in some sense the complement of capacitors. Like Capacitors, inductors have a memory, i.e. they store energy. While a capacitor stores energy in its electric field, an inductor stores energy in its magnetic field. The fundamental formula for an inductor is:

$$v = L \, di/dt$$

Where L is the inductance of the inductor as measured in units of Henrys, symbol "H". An inductor behaves like a short circuit for a steady current, and like an open circuit for a very rapidly changing current. Thus, opposite from the capacitor, when analyzing the DC case replace all inductors with short circuits, and for very rapid changes replace all inductors with open circuits.

Physically inductors are most commonly created with loops of wire, much as the abstract symbol (see figure 9-1) suggests.

Inductors in series add, inductors in parallel combine with the familiar formula:

$$L_{equivalent} = L_1 \parallel L_2 = (L_1 L_2) / (L_1 + L_2)$$

When you have a single inductor and a resistor, the response to a sudden change is an exponential function just as with the case with the capacitor, only the time constant $\tau = L/R$.

Like capacitors and resistors, ideal inductors are linear circuit elements. This means that you can use all the same rules of superposition etc. for circuits involving inductors.

Warning: whenever you have more than one energy storage element the analysis becomes complicated, well beyond the scope of this text. Energy can transfer back and forth from capacitor to capacitor, or from inductor to capacitor etc., you can get resonances and oscillations and all sorts of other strange effects.

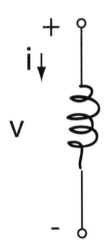

Figure 9-1
Inductor Symbol.

45

Find an engineer or use a computerized circuit-analysis program. The only exception is when you can trivially combine elements in series or in parallel. Note than you cannot combine both a capacitor *and* an inductor in series or parallel in any simple way.

Inductors are of relatively little interest to most biologists. For technical reasons inductors are not as easy to make as capacitors and they tend to be more sensitive to noise, thus instrumentation is usually loaded with capacitors but has few if any inductors. Also, while capacitative effects are strong for biological tissues, inductive effects are so weak in comparison that they are usually ignored.

Any time a current flows in a conductor it has a certain amount of "self-inductance", i.e., the current generates a magnetic field that can be modeled as an inductor. For biological tissues this *self-inductance* is so small as to be almost always physiologically irrelevant. However, the induced magnetic fields do exist, and very sensitive super-conducting magnetometers are sometimes used to map current flow in (for example) the axons and dendrites in the brain. As of this writing the possibility that similar effects might be recorded via techniques of functional Magnetic Resonance Imaging (fMRI) is being investigated.

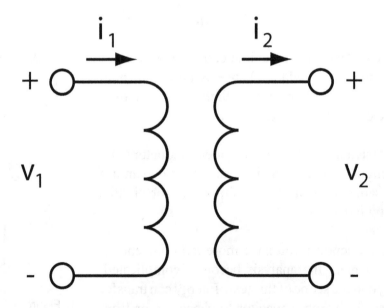

Figure 9-2. Transformer circuit diagram.

Speaking of MRI, this is a technology that makes use of large coils of wire in strong magnetic and Radio-Frequency (RF) fields, and inductive effects are LARGE. However, the operation of MRI devices exists in a regime where the lumped circuit element models break down – be warned that MRI physics is hard and not covered in this text.

Inductive effects are also present in transformers. A transformer is essentially two inductors that are coupled together via their magnetic fields. A transformer takes an AC current and, depending upon the ratio of turns of wire in the two coils that make it up, can either increase the voltage at the cost of decreased current or decrease the voltage while yielding increased current. This only works for AC currents: a DC current will burn out the transformer, because to a DC current the coil of wire in a transformer looks like a short circuit.

Transformers are sometimes found in high-frequency circuitry to match antennas and transmission lines. Transformers are very commonly found in electrical power supplies, and because of their use of magnetic fields to couple the two coils they are a strong source of electrical interference.

Keep all power transformers as far away from where you are making sensitive measurements as possible!

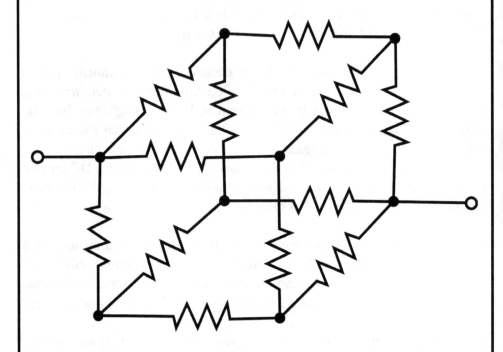

Imagine that you have a bunch of resistors, each of which is 1000 Ω, and they are connected along the edges of a cube. What is the total resistance as seen from the two external terminals (represented by open circles)?

HINTS: Because the resistors are the same, you can do this easily using symmetry. Remember that nodes that are of the same electrical potential can be connected together. That should simplify the circuit, and you can then use resistors in series and resistors in parallel to solve.

Note that if the resistors were of different values, the problem would not be so simple! And if you really want to, you can always get some resistors, connect them up as shown, and measure the resistance across the terminals using a multimeter. Checking your results against physical reality is often a good idea!

10. Diodes/Nonlinear Elements

Not everything in life is linear. Some of the most interesting and important phenomena are nonlinear. Unfortunately there are no simple rules that you can use for all nonlinear systems, you basically have to perform a new analysis for each nonlinear circuit. Remember, if a circuit has a single non-linear element in it, then even if everything else in the circuit is linear, the operation of the entire circuit must be considered to be potentially nonlinear. Analysis of nonlinear systems is a topic of considerable ongoing effort. At present, the most general-purpose method is to use computerized systems that use various successive-approximation techniques to "brute-force" the answer.

The simplest non-linear circuit element is the *ideal diode*. A diode can be thought of as a one-way valve, that is a perfect conductor to current flowing in one direction and a perfect insulator to current flowing in the other direction. An ideal diode can also be characterized by its current-voltage relationship (see figure 10-1). Some texts distinguish ideal from real diodes by placing rectangular boxes around the symbol, as in this figure, but this is not consistent.

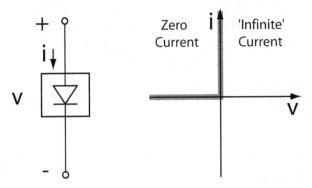

Figure 10-1. Ideal diode, symbol (left) and current-voltage relationship (right).

For negative ("reverse") voltages the ideal diode passes no current. For positive ("forward") voltages the diode can pass unlimited current – that is, in the forward direction the diode acts like a wire, the voltage across it must be zero, and any value of current can be carried.

Like the linear resistor, diodes do not store energy, they have no memory. Thus, an analysis of a circuit involving only diodes and resistors does not have to take time into explicit consideration.

Real diodes used for electronic circuits deviate from the ideal quite a lot more than real resistors do. The deviations vary with the precise method of construction, but very roughly real diodes are characterized by a *threshold voltage* ("forward bias") required to turn the diode "on" in the forward direction (varies from 0.2 to 0.7 volts, depending on what the diode is made of), and a *reverse voltage breakdown* where a diode that is in the nonconducting state suddenly starts conducting when the voltage gets too negative.

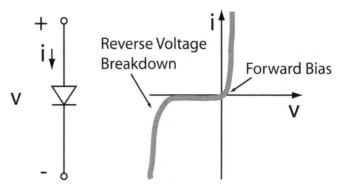

Figure 10-2. Real semiconductor diode i-v characteristics.

Diodes designed for use in circuits can be optimized for signal processing, these are usually small glass cylinders with two wire leads and a black band around one end (corresponds with the bar on the circuit diagram). Some diodes are designed for use in power circuits, these are typically made of a black plastic with a gray ring around one end and depending upon the maximum current and voltage ratings these diodes can be rather large.

Other diodes have precisely calibrated reverse voltage breakdowns, these are called Zener diodes and are used in circuits as a source of constant voltage ("voltage regulator").

Diodes are used in too many different kinds of circuits to list here. Any decent electronics handbook will give you dozens to choose from.

Biophysically many ion channels have a different conductance in the forward and reverse directions. Such ion channels are said to be *rectifying.*

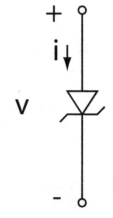

Figure 10-3. Zener diode.

A rectifying ion channel is never a good match for an ideal diode, but with the correct extra elements ideal diodes can be used as parts of a model that gives a pretty good fit (see figure 10-4). However, a careful analysis will more typically use computerized modeling techniques where you specify the exact form of the i-v curve.

Any ion channel or electrode contact that does not rectify or have any other strong nonlinearities is sometimes called an *Ohmic* channel or contact, to signify that it can be modeled more-or-less accurately as a linear resistor that obeys Ohm's law.

There are many other nonlinear circuit elements in use, most notably transistors. The standard transistor has not two but three leads coming out of it, and is highly nonlinear. There are two good things to know about transistors. The first is that transistors can use a small current to control ("gate", "modulate") a larger current, that is, a transistor can amplify a signal. The second is that anything a biologist could ever want has already been built and perfected by highly trained engineers, so you don't need to know anything at all about these little beasts, which is a good thing, as using them in a circuit design is not for the faint of heart.

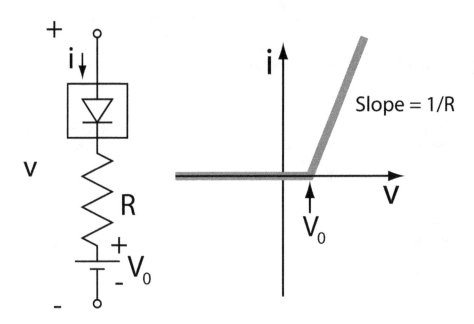

Figure 10-4. Modeling a nonlinearity using simpler ideal elements.

Electronics Apocrypha – the BNC connectors and coaxial cables. There are more different types of electrical connectors than any one text could ever cover, and more different types get developed every day. However, perhaps the most common connector on electronic equipment is the "BNC" connector, which is usually hooked up to a coaxial cable. The right image shows two BNC panel-mount connectors on a piece of electrical equipment, and the left image shows the end of a coaxial cable with a BNC connector attached. The BNC allegedly stands for "Bayonet Neill-Concelman", after the bayonet-type mechanism and two of its inventors. Some sources, however, claim that BNC stands for either the British Naval Commission, British Naval Connector, or Bayonet Nut Connector.

This is a good all purpose connector, but it has been optimized for use in relatively high-frequency systems (typically up to 3 gigahertz). The cable consists of two conductors, a single wire running down the middle, and a cylindrical conductor just inside the external insulation. The two conductors are separated by another layer of insulation. This works well for high-frequency signals, and provides some inbuilt shielding against noise, but there are issues for using them in a biology lab that you should be aware of.

-> Coaxial cable has a high capacitance relative to plain wires. You should never feed a weak electrical signal directly into an ordinary coaxial cable. You should first connect the signal to a preamplifier, or buffer, using short simple wires, THEN use coaxial cables to carry the signal a longer distance.

-> Coaxial cables don't always take well to being frequently plugged in and unplugged – and yet, this happens a lot in a biology lab. These connectors can readily break or become flakey, and if you take them for granted, you can waste a lot of time trying to figure this out. Destroy and throw away all suspect cables. Do not try and save money by making your own: buy commercially assembled ones. Trust me.

-> Most BNC connectors ground the outer connector in the cable to the metal chassis and ground of the instrument that they are connected to. This can cause ground loops or other issues if you forget about that.

11. OP-AMPS

It is a good rule of thumb that biologists should never try to build instrumentation from scratch. Biologists have more than enough to occupy their minds as it is, and nearly anything a biologist could want is already made and perfected. Don't re-invent the wheel!

Trying to save money by building your own equipment is also generally foolish. Remember Gawne's Law of the Lab: *everything is hard*. A circuit may look simple on paper, and only need $36.75 of electronic parts from Radio Shack, but when you take into account making a box for it, creating a stable and safe power supply, dealing with changes in temperature and humidity, making it resistant to noise and electrical interference, making it mechanically rugged, dealing with overvoltages and accidentally mis-connected inputs and wires, etc.etc.etc.etc., it's almost always just not worth the effort. Even if you have ready access to a talented engineer, let them work on new things and on developing interfaces between ready-built pieces of equipment, rather than waste their time on a false economy. Really!

However, there are times when you do need to cobble together something that hasn't been made, and more likely, there are times when you need to be able to understand a schematic circuit diagram that someone else has come up with. The single most useful piece of circuitry (other than resistors, capacitors, batteries, and wires!) is the *operational amplifier*, or "Op-Amp". In essence, the op-amp has encapsulated many transistors and, using a sophisticated design, come up with a building block circuit element that appears simple from the outside, and which by hooking it up in different ways can be made to perform an amazing number of operations.

An op-amp has a minimum of five terminals (see figure 11-1). It has a positive and a negative voltage *supply* (V+ and V- in the diagram, these are often +15 and -15 volts), an inverting input, a non-inverting input (awkward but there is no appropriate English antonym for inverting), and an output.

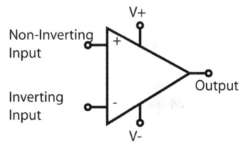

Figure 11-1. Op-Amp Symbol.

In most circuit diagrams the power supply connections are omitted, but they are always there. An op-amp requires external power for its circuitry; it is an *active* device (as opposed to resistors, capacitors, inductors and diodes that are always what they claim to be, battery or no, and are called *passive* devices). There are also sometimes additional connections for fine-tuning the op-amp.

The op-amp is a very high-gain *differential amplifier*. The output voltage is

$$v_{OUTPUT} = A(v_+ - v_-)$$

That is, the op-amp amplifies the difference in voltage between the two input terminals. Figure 11-2 illustrates the input-output relationship of a typical OP-AMP. v_{OUTPUT} cannot, however, exceed the positive voltage supply or decline below the negative voltage supply, if v_{OUTPUT} tries to exceed this limit and fails the amplifier is said to *saturate* and v_{OUTPUT} remains stuck at the value of the supply voltage that it tried to exceed. For most signal-processing applications you will want to make sure that the output voltage stays within the range spanned by the positive and negative supplies – this is the linear range of operation. Also, the voltages at v+ and v- have to stay within this range as well, both to ensure linear operation, and to avoid damaging the device.

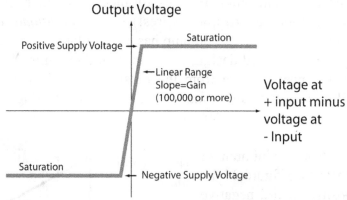

Figure 11-2. Op-Amp Input-Output Relationships.

A is the *open-loop gain* of the amplifier. The gain is usually very large: 100,000 or more, and it varies widely from one op-amp to the next, and it also varies a lot with temperature. So what is the point of having a poorly-controlled amplifier with such an outrageous amount of gain? We can crudely think of it in this way: engineers like tradeoffs. Speed vs. power, strength vs. lightness, you get the idea. The tremendous gain of an op-amp is a resource that can be selectively traded back in exchange for other features, such as stability or precision.

The simplest op-amp circuit is the *voltage follower*, or unity-gain amplifier (see figure 11-3). Feedback from the output to the inverting input forces the output to follow the non-inverting input. Even though the voltage gain is 1 (just like a wire!) this is an extremely useful circuit because the input impedance can be very high (depends upon the specific op-amp circuit) and the output impedance is usually low, this is an excellent and commonly used *buffer* or *pre-amplifier* or *headstage* circuit.

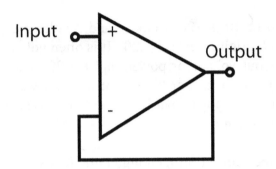

Figure 11-3. Voltage follower.

Important point: the precision of the voltage follower is to a great extent unaffected by changes in the open loop gain. Think about it this way: if the open-loop gain of the op-amp is 1,000,000, then the voltage follower will tend to get the output to within $1/1,000,000^{th}$ of the + input. If the open-loop gain of the op-amp is 2,000,000, then the voltage follower will tend to get the output to within $1/2,000,000^{th}$ of the + input. This is a considerable change in the relative sizes of the errors, but in both cases the absolute overall gain of the circuit with feedback is very close to 1.0.

Circuits to perform addition, subtraction, different levels of amplification and nearly anything else can be found in any standard electronics text, if needed.

Decibels

Decibels are a tricky thing. There used to be a unit called the "bel", and the decibel was a tenth of a bel, but nobody uses "bels" anymore, so the decibel ("dB") is the prime unit. Originally the decibel was defined in terms of power gain (power amplification), as:

Power gain (Decibels) $= 10 \log_{10} (P_{OUT}/P_{IN})$

However, in most (not all!) cases the decibel is used as a ratio of amplitudes, and not (as originally intended) as a ratio of power. As we recall, if we are dealing with resistors, power is proportional to the square of the voltage, so in this case:

Voltage gain (Decibels) $= 20 \log_{10} (v_{OUT}/v_{IN})$

Hence a voltage gain of 10 is +20 decibels, and a voltage gain of 0.10 (attenuates to one-tenth original level) is –20. It is often not clear whether someone is talking about voltage or power, and the 20 factor is frankly weird, but we are stuck with it. For the rest of this text we will refer to voltage (absolute signal level) gain, but be warned, not every reference to decibels uses this convention.

For all its faults, decibels are a useful enough unit. It is a dimensionless, logarithmic measure of the ratio of two quantities. An amplification of 20 dB is a gain of 10, an amplification of 40 dB is a gain of 100, an amplification of 60 dB is a gain of 1000, and so on. Also, as a logarithmic unit, if you feed a signal through two amplifiers in series, you just add the gain in dB. For example, imagine that you have two amplifiers each with a gain of 10x (20dB) and the output of the first feeds into the input of the second. The overall gain of both combined is 10x10 = 100, or 20dB + 20dB = 40 dB.

If one signal is twice as large as another, the ratio is 6dB. Often in filtering signals we will talk of the "-3 dB point", which is the point where the signal is reduced by 0.707 of its original value. 0 dB means unity gain, i.e., the input equals the output.

Decibels are sometimes used as logarithmic measures of absolute physical quantities, but in this case there is always an assumed physical reference standard (for the acoustic measure of decibels it's a previously defined sound pressure level) that it is compared to.

Chapter 12. Linear Systems and the Impulse Response

There is an old joke about the drunk who loses his house-keys, and is found looking for them under a lamppost. When asked where he lost his keys, he replies near the door. When asked why is he looking under the lamppost when the door where he lost the keys is far away, the drunk replies: "Because there is more light here."

This is not a bad strategy. Even if the odds are low that the keys bounced over to the lamppost, if that's the only place you can see anything then you might as well start looking there. Linear systems are the only systems for which we have a well-developed set of general, guaranteed-to-always-work analysis tools. Instrumentation is usually carefully designed to be as linear as possible. Many real systems of interest are more-or-less linear, at least for restricted ranges. Even if a system is significantly nonlinear, a linear analysis will usually work for a small enough range of variation ("small-signal analysis"). And ultimately, how can we determine if a system is nonlinear if we don't first know what a linear system is?

We have already encountered some important aspects of linear systems. The first is superposition, which for electric circuits means that you can analyze a circuit with multiple voltage and current sources by calculating the response of the circuit to each component in turn. Other circuit tricks are being able to create equivalent circuits, or realizing that the response of a linear network with a single capacitor to a sudden change is always some variant of $e^{-t/\tau}$, where τ is C times the equivalent resistance as seen from the capacitor.

But what if the input to a linear system are neither constant, nor abrupt step changes? What if the input to a circuit is a complex function of time, for example, a voltage that represents an audio signal? There are two main approaches that can be used. The first is a fairly brute-force technique that uses the impulse response and the operation of convolution. The second decomposes a signal into a sum of sine-waves of different frequencies and phases and uses operations in the frequency-domain to analyze a system. Both approaches have their uses, and while the full range and derivation of these techniques is beyond this text, an understanding of the basics is important for any biologist interested in electronic instrumentation or the electrical properties of tissues.

Continuous-time and discrete-time signals

Figure 12-1 shows a discrete-time (top) and a continuous-time (bottom) signal. So far we have been dealing only with continuous-time systems. For example, sin(t) is continuously defined for all points in time. However, the advent of digital computers has made discrete time signals increasingly important. A computer cannot store an infinite number of voltage values to represent the infinite number of voltage values for a continuously-defined signal. It is standard practice in digital signal processing to sample a continuous waveform at some regular, rapid rate, and to use these discrete samples to approximate a continuous signal. The currently popular compact disc technology for storing music samples the audio signal at about 48,000 samples/second, a rate fast enough to accurately represent virtually all of the information carried in a continuous-time audio signal.

The discrete time signal in the top of figure 12-1 is illustrated in a "stem" plot to emphasize that the function is only defined at specific points. It is standard to plot discrete-time functions by connecting the dots, which gives a more intuitive sense for how a function is changing over time, but which obscures the discrete nature of the function.

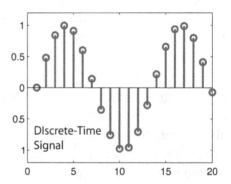

There are subtleties involved in using a discrete-time signal to represent a continuous-time one. We often use a digital computer to record a continuous-time waveform, and this requires creating a discrete-time waveform by sampling a continuous-time one at regular intervals. If the sampling rate is slow compared to the variations in the continuous signal, then the discrete-time signal will be appear to be erratic and even random, a condition known as *aliasing* (see figure 12-2).

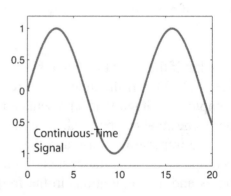

Figure 12-1. Continuous-Time and Discrete-Time Signals.

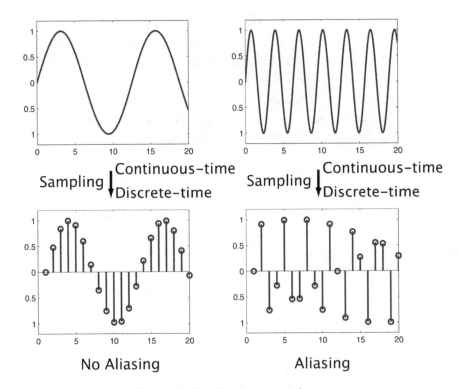

Figure 12-2. Aliasing example.

For processing discrete-time signals within a standard digital computer numbers are represented in floating-point format, so we can regard the discrete-time signals as being continuous in amplitude. However, for signals that are created by sampling directly from an external continuously-variable ("analog") signal, there are other issues. A digital computer system accesses an external voltage via an Analog-to-Digital Converter (ADC, or A/D Converter). These converters <u>always</u> have a limited number of specific input levels that they can recognize. For example, a 16-bit analog-to-digital converter can represent $2^{16} = 65536$ different voltage levels. If the converter is set to span a range of input voltages from 0 to +10 volts, the minimum change that you can detect would be $10*(1/65536) =$ about 0.00015 volts.

If the variation in a part of the analog signal is so small that it all falls within the range of one step change of the ADC, then you are obviously not going to be able to see any changes in the continuous signal reflected in changes in the discrete-time signal. In addition, if the variation in the input signal is too large, you will not be able to respond to the signal when it goes out of range. Digitizing an analog signal requires a balance: you want to use as much of the range of the A/D converter as possible, to maximize your sensitivity to small changes in the input signal, but not make the range so small that you miss part of the signal entirely. Digitally-based signals are thus often discrete in two respects: in time and in amplitude. It is obligate that these signals are discrete in time: that they are discrete in amplitude is due to the vagaries of the computer systems on which they are commonly implemented. In the purely mathematical treatment of discrete-time signals you can use real numbers for the amplitude and thus the discrete in amplitude condition does not apply.

While discrete-time signals are often used to represent a physical continuous signal, they are legitimate signals in their own right and have a fully developed mathematics that does not depend upon there being any corresponding continuous signal. When you sample a signal via an ADC, you specify a sampling rate, for example, 1 millisecond per sample. In this case the first discrete sample represents the signal at time t=0, the second sample represents the signal at time t=0.001 seconds, etc.

However, for the discrete signal itself the sampling rate is irrelevant. The discrete signal is defined on its own terms, and mathematical operations need only specify the discrete time epochs (the "index"). The index is most commonly represented as being an integer. Any relationship between an actual physical signal and a discrete-time signal is wholly at the discretion of the user, and is not intrinsic to the discrete signal itself. Hence, for a signal that we consider to represent millisecond-separated samples of a continuous signal, the discrete signal is itself still represented as (for example) x[0], x[1], x[2], …. It is up to us to know that, in this case, each increment of the index corresponds to 0.001 seconds.

It is trivial but bears repeating that while we most commonly discuss signals as a function of time, they can just as easily be a function of position as well. The same sorts of analysis can be done in both cases.

Impulse Response

This important concept is most easily appreciated in the discrete time case. Consider a linear time-invariant (LTI) system. This is a linear system whose parameters do not vary with time. The signals vary with time, but the system itself does not. You input a signal in one side, and you get another signal out the other. An example of a linear time-invariant system is a stereo amplifier with the volume knob locked in position. A linear time-varying system is a stereo amplifier where the volume knob is slowly

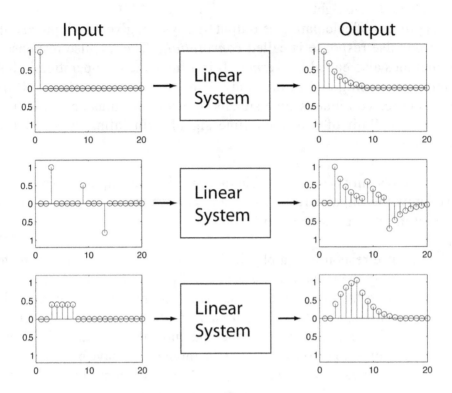

Figure 12-3. Discrete-Time Impulse Response.

moved back and forth. People feel obligated to specify the "time invariant" part in basic texts (where everybody is trying to be super-precise), but the usual assumption is that all linear systems are time-invariant unless specifically marked otherwise.

Now consider the response of a system to a discrete-time signal that consists of a single value of one at time = 0, and which is zero everywhere else (see figure 12-3, top row). This simple signal is known as the unit impulse. If we use the unit impulse as input to a system, we will get some other signal out. Here is the critical part: because linear systems obey scaling and superposition, we have now completely characterized the response of this system and can predict the output for ANY input. That's because any input can be considered as the sum of a series of unit impulses that are appropriately shifted in time and scaled in amplitude. We then build up the output by adding together the separate responses to each unit pulse one at a time. If one part of the input is a pulse at time t=2, and the magnitude is –5, then the output in response to that one part of the input is the response to the unit impulse scaled by –5, and shifted two places to the right.

This process of calculating an output of a system given the input and the unit impulse response is called *convolution*. The impulse response is sometimes also called the *kernel*. It is a fundamental operation in both signal processing and in the study of many physical systems. It can be done for continuous-time signals by using calculus and essentially taking the limit of a discrete-time signal with infinitely small time increments.

A linear system is completely characterized by its impulse response. If we know the impulse response of a linear system we can calculate the output in response to any input.

The impulse response of a physical system can often be approximated by measuring the response to a very brief pulse. Techniques also exist for extracting the impulse response given the input and the output, this is part of the broad field of *system identification* techniques. Finally, if you know the impulse response, and you know the output, sophisticated algorithms exist to reconstruct what the input had to have been, a process known as *inverse convolution*. The inverse of convolution is not a simple process, and unlike regular convolution it is does not always work (it often fails due to stability concerns), but it can be a handy tool to have access to.

13. Sinewaves and Filters

In principle you can do everything in the time domain. Signals can be completely specified as a function of time, and if you know the impulse response of a linear system then you know everything there is to know about that system. However, many operations are technically hard to perform or yield little insight when done as a function of time. One powerful approach to analyzing linear systems is to decompose a signal in the time domain into the sum of sine waves of different magnitudes, frequencies and phases, perform operations in what is called the *frequency domain*, and then transform back to time.

It's sort of how people used logarithms in the days before pocket calculators: to multiply two numbers you take their logarithms, add them, and then calculate the inverse logarithm. In a similar way you can transform signals into a sum of sine waves, then perform operations that are hard/nonintuitive in the time domain but simple when considered in terms of frequency, and then convert back to time.

The mathematics of this approach is incredibly developed, and a full treatment is a large part of what an engineering education consists of. However, much of the basics can be understood and used without too much difficulty.

Figure 13-1 illustrates a garden-variety sine wave, like you learned from high-school trigonometry. A sine wave has three parameters: a magnitude, a frequency, and a phase. Engineers often talk about *angular frequency* ω, measured in units of radians per second, because it makes the math cleaner. Angular frequency is related to traditional frequency "f" (cycles per second) with the formula $\omega = 2\pi f$. Frequency "f" and the period of a sine wave "T" (the duration of a single period) are reciprocals of each other, i.e., $f=1/T$.

Sine waves are important for several reasons. They are the solutions of several classes of differential equations, they appear in many natural phenomena, and they have unique properties when it comes to linear systems. In particular, for any linear system, if you input a sine wave, the output is guaranteed to also be a sine wave of the same frequency. For sine waves, only the magnitude and phase can be changed by a linear system.

If the input to a system is a sine wave, and the output is anything other than a sine wave, then the system is not linear.

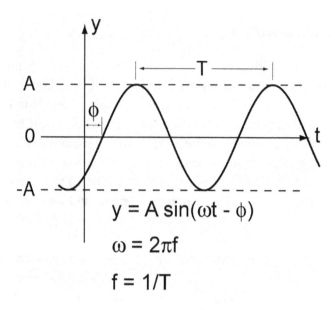

$$y = A \sin(\omega t - \phi)$$
$$\omega = 2\pi f$$
$$f = 1/T$$

Figure 13-1. Basic sine-wave.

Fourier Transform

Another useful property of sine waves is that any function can be decomposed into the sum of a series of sine waves of differing magnitude and phases. Thus, you take a signal that is a function of time, and now represent the same signal as the coefficients of a series of sine waves of differing frequencies and phases. We say that we can transform a signal in the *time domain* into a signal in the *frequency domain*.

Consider the frequency-domain representation of a square wave (see figure 13-2). A "square wave" is a repeating signal that alternates equally between two values. Like a sine wave, an ideal square wave is defined for all time, i.e., from $-\infty$ to $+\infty$. The formula for representing a square wave as a sum of sine waves is:

$$y(t) = \text{sum } (n=1,3,5,7,\ldots) \ (4/n\pi) \sin(n\omega_0 t)$$

Where $\omega_0 = 2\pi/T$, and T is the period of the square wave. The value of the sine wave for each value of n is sometimes referred to as a *harmonic* of the square wave. (The first harmonic n=1 is known to musicians as the *fundamental* of the square wave.) As you add more and more higher harmonics, the sum of the sine waves more closely approximates a true square wave. For signals that have an abrupt step in time, it takes an infinite number of harmonics to exactly match the desired signal. Fortunately,

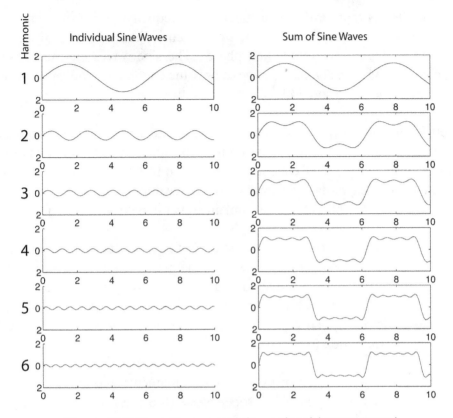

Figure 13-2. Example representing a signal (square-wave)
as a sum of sine waves.

as the harmonics increase in number the magnitude of each succeeding harmonic becomes progressively less, so that we can get a pretty good fit with a reasonable number of harmonics.

It is a truism that time-domain signals with sudden changes require sine waves that span a wide range of frequencies to represent them. In engineering jargon, we would say that signals with abrupt changes have broad frequency spectrums. Conversely, a signal that is a pure sine wave has a very narrow frequency spectrum (just the single frequency of the sine wave).

We go from time to frequency using the *Fourier transform*, and from frequency to time using the *inverse Fourier transform*. Often the Fourier transform will be referred to as the Fast Fourier Transform, or FFT. The FFT is merely one specific method of performing a Fourier transform on a digital computer, but somehow it has become a commonly used synonym. Important: a signal that is represented as a function of time consists of a series of numbers, one for the value of the signal at each point in time.

A signal that is represented as a function of frequency is represented as a series of numbers (coefficients) each of which is the magnitude of a sine wave of a specific magnitude and phase. The sine waves themselves do not change from one signal to the next, only the coefficients indicating how large each sine wave should be for that specific signal.

There is a uniqueness proof, which we won't go into here, stating that you can transform a signal in the time domain into one and only one corresponding set of coefficients in the frequency domain, and that you are guaranteed to be able to go backwards and from the frequency domain representation of a signal and recover the exact time-domain signal that you started with.

Phase is often neglected when compared to the magnitude of the Fourier transform, but it plays a critical role. Get the magnitude of the sine waves correct but scramble the phases and you cannot recover the original signal. Additionally, filters and other signal processing systems that do not alter the magnitude of the sine waves but that shift their relative phases can add considerable distortion to a signal.

Figure 13-3 is an example of the first six components that make a square wave, shown added together in proper phase (top row) and also with the phases scrambled (other rows). Only when the sine waves are all at the correct phase do they sum together to yield a proper square wave.

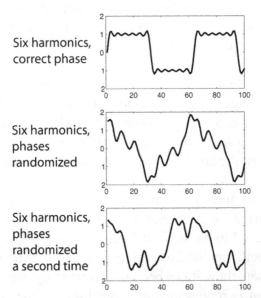

Figure 13-3. The importance of phase.

While the principles of frequency analysis are simple, there are a lot of subtle points. We will try and skirt around the difficult issues and concentrate on those aspects of frequency analysis that are of direct relevance to biologists.

The decomposition of a time-varying signal into sinusoids is rarely as simple in practice as with our simple square-wave example. In particular, the conversion of real physical signals usually involves sampling the signal for a fixed period of time, and not from positive to negative infinity. Digital computer techniques can only handle a finite number of time samples and a finite number of frequencies, which can lead to various distortions and a 'smearing' of the frequency components.

Additionally, in order to correctly preserve the information about phase, the mathematically precise Fourier transform uses complex numbers, as well as positive and negative frequency (!). Most typically we will look at the *frequency spectrum*, which is just the magnitude (absolute value) of the Fourier transform. This is a useful method of representing a signal, but remember that by leaving out spatial phase information the representation is incomplete.

Sometimes you will encounter a *power spectrum*, which is the point-by-point square of the frequency spectrum. That's because power in many physical systems (not just electronics) is often proportional to the square of some variable.

In the examples given in figure 13-4, the graphs on the left side represent a signal as a function of time, and the graphs on the right the same signals are represented as a function of frequency. A sine wave gives a peak at one frequency, smeared out a bit because we only sampled for a finite period of time. If we had sampled for longer the peak in the frequency plot would have become sharper, and for a shorter period of time the peak would have smeared out further. As expected a higher frequency sine wave yields a peak at a higher frequency.

In the bottom row of figure 13-4 we see that for a constant ("DC") signal the frequency spectrum is everywhere zero except at "zero frequency", i.e., a constant level. Think of it as the ultimate in low frequencies!

In the top row of figure 13-5 the signal is so-called "white noise", where the value of each point in time is random and uncorrelated with any preceding

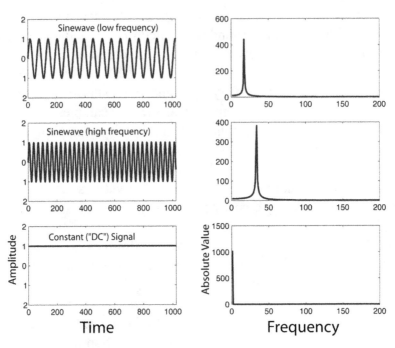

Figure 13-4. Examples of fourier transforms of some common signals.

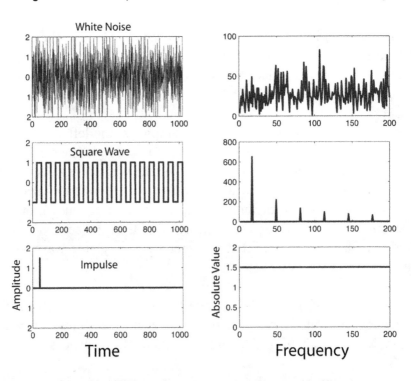

Figure 13-5. More examples of the Fourier transform.

or following points. The frequency spectrum of this signal is noisy but basically constant as a function of frequency. Hence the adjective "white", because like white light, white noise has an equal amount of energy per unit frequency.

Sometimes you will hear reference to "pink noise". This is similar to white noise except it has equal energy per octave (i.e, per doubling of the frequency), and translated to visible light would result in a color that is slightly pinkish.

The middle row of figure 13-5 is a square wave, and in this case we see the first six harmonics in the frequency domain. There are an infinite number of harmonics, of steadily increasing frequency and decreasing magnitude, we have truncated the frequency scale to a finite length.

In the bottom row of figure 13-5 we have an impulse, i.e, an infinitely narrow pulse. The frequency spectrum is flat (constant) for all frequencies – up to infinity, if we bothered to draw that far. In the real world you will never encounter an infinitely abrupt pulse, all real pulses/transients have some degree of smoothness to them, and a correspondingly limited range of frequencies. Nevertheless, signals with abrupt changes will have sine wave components that span a broad range of frequencies.

The single most useful part of the Fourier transform is when you encounter signals that overlap in time but that do not overlap in frequency. In this case you can trivially use a frequency-domain representation of a signal to separate these two signals, when doing the same thing in time, while always technically possible, would be very hard.

Simple RC Filters
For the two filters illustrated in figure 13-6 v_{input} is the input, and v_{output} is

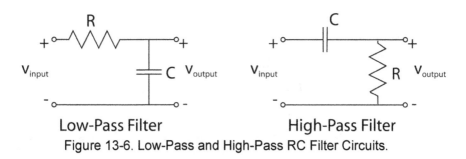

Low-Pass Filter   High-Pass Filter
Figure 13-6. Low-Pass and High-Pass RC Filter Circuits.

the output. It should be apparent that these two circuits are really the same, just a resistor and a capacitor in series. For the low-pass filter we take as output the voltage across the capacitor, and for the high-pass filter we take the voltage across the resistor.

We already know what to do for the DC (steady-state) condition: remove the capacitor. The filter at left then passes the voltage from input to output unchanged, and the filter at right blocks the signal completely. We also know that for extremely rapid changes we replace the capacitor with a wire: now the circuit at left blocks the signal (C shorts out the terminals forcing v_{output} to be zero) and the circuit at right passes the voltage though without any attenuation.

We also know what to do for step changes in the input voltage: figure out the starting and ending conditions, and connect them with an exponential function of the form $e^{-t/\tau}$ or $(1-e^{-t/\tau})$ scaled and adjusted to fit, where the time constant $\tau = RC$.

But what about signals that vary as a more general function of time? We could write out the differential equation for a capacitor and solve for that, and there are times where that is necessary, but it's hard work and not very intuitive. An easy way to understand the operation of these circuits is to look at the ratio of the output to the input voltages as a function of frequency, in what are called Bode plots.

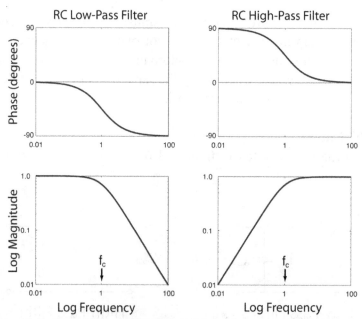

Figure 13-7. Bode Plots for Simple RC Low-Pass and High-Pass Filters.

The magnitude and phase of an output sine wave, relative to that of an input sine wave, is plotted in figure 13-7 as a function of frequency for the low-pass (left half) and high-pass (right half) circuits. Frequency is plotted on a logarithmic scale, centered at the "corner" frequency, f_c. The corner frequency is determined as $f_c = 1/2\pi RC$. For the low-pass filter the magnitude of the response is near 1.0 for frequencies below the corner frequency, thus sine waves with frequencies below f_c are passed through with little attenuation. As frequency approaches the corner frequency, the signal is increasingly attenuated, until at f_c the signal is down by 3 dB, or about 0.707 of the original value. This is the *3 db point*, it is the most common (essentially arbitrary) standard for defining when a filter switches from passing a signal to rejecting it. Note that this is the point where the power in the signal is halved.

As frequency is increased, attenuation steadily increases. On a log-log plot the magnitude approaches a straight line, which is either 20 dB per decade or 6 dB per octave (doubling of frequency). Thus, increasing the frequency by ten times attenuates the input signal by an additional factor of ten.

The phase plot for the low-pass filter has an identical logarithmic scale for frequency but a linear scale for phase. At low frequencies the phase shift is near zero, and the phase of a sine wave as input is maintained in the output. Thus signals below f_c are allowed to pass though this filter with minimal changes in amplitude or phase, as the input frequencies pass through f_c the amplitude of the output starts to decrease and frequency-dependent phase shifts become significant. High above f_c there is very little signal that makes it out.

The high-pass filter is really the same only with the effect of changing input frequency inverted.

These are two of the most useful circuits that you will ever see. Even extremely sophisticated digital filters often still need circuits like these to pre-filter the inputs to an appropriate frequency band, for example to avoid aliasing But be warned: these filters only have these characteristics if they are driven by an ideal voltage source and connected to a load with infinite (or at least much larger than R) input resistance.

As illustrated in the middle row of figure 13-8, if you connect a simple RC filter to other circuits that have resistances on the order of the R used in these circuits, the resistances of the other circuits will modify the value of the cutoff frequency! You can isolate the circuit from being influenced by what it is hooked up to by using a pair of voltage-follower circuits (bottom row of figure 13-8), which works great in isolating the function of the circuit from changes in what it is connected to, but now the circuit needs a power supply to run the op-amps, and also requires that all voltage signals lie within the range of the power supplies of the op-amps, and in addition it only works up to about 100 Khz or so beyond which the op-amps themselves start to act as low-pass filters.

It is typical to use a single capacitor in series to "AC-Couple", to block low frequencies. When you set an oscilloscope to AC-Couple on its inputs, all

Low-Pass Filter Works as Advertised

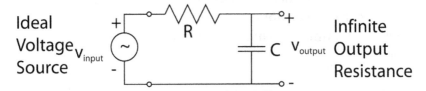

Filter Changed by Connected Circuits!

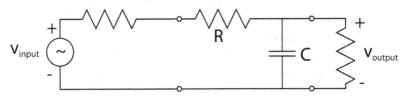

Filter Isolated via Op-Amp Buffers!

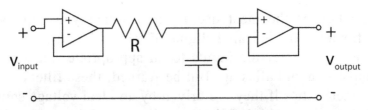

Figure 13-8. The Perils and Salvation of Hooking up RC Filters to Other Circuits!

you are doing is switching a capacitor in series with the input. Remember, though, that the exact frequency at which a "blocking capacitor" starts to attenuate depends upon the resistances (impedances) of the equipment that it is hooked up to.

There are a variety of different types of filters, the major classes are shown in figure 13-9. The band-pass filter passes only a restricted range of filters, the notch filter blocks only a narrow band of frequencies (useful if you have a lot of noise of one narrow frequency), and the seemingly useless all-pass filter is sometimes seen, it typically has phase-shift properties needed by some specialized applications.

Higher-Order Filters

More sophisticated designs can be created using multiple independent (i.e., cannot be combined via series or parallel) capacitors and inductors, or capacitors and op-amps. These more sophisticated filters are typically ranked according to how many *poles* they have. This term comes from the analysis of time-varying circuits that uses complex numbers: for now think of the number of poles as being sort of like the number of cylinders in an automobile engine, with more being potentially more powerful at the downside of being more expensive and harder to maintain.

Even with the same number of poles, there are different classes of filters that vary in how sharply they switch from passing to rejecting a signal.

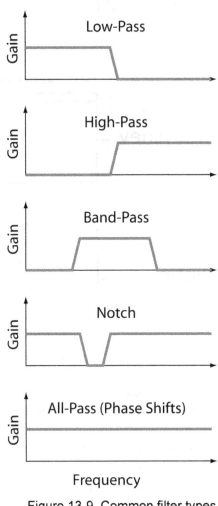

Figure 13-9. Common filter types.

73

Figure 13-10 shows the frequency response (left side) and the time-domain response to a brief rectangular pulse (right side) for Bessel, Chebyshev and Elliptic low-pass filters. The Elliptic and Chebyshev classes have sharper cutoffs in the frequency domain, but at the expense of introducing significant ripple and distortion in the time domain. They "ring" in response to abrupt changes in a signal, the same way that a car without shock absorbers bounces up and down several times when it hits a bump. When recording abrupt changes in an electrical signal, such as an action potential in a neuron, this "ringing" can add extra peaks to the data that do not actually exist.

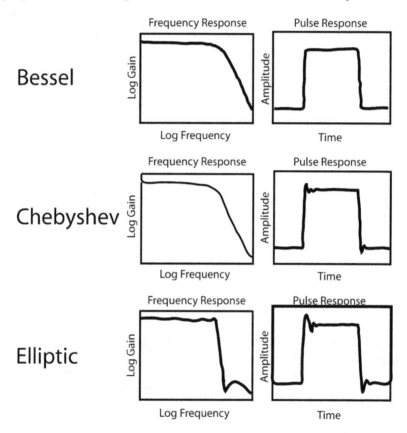

Figure 13-10. Distortions of a waveform by different classes of filters.

The Bessel filter does not have as sharp a cutoff in the frequency domain, but it does not introduce spurious ripples or extra pulses in the time domain. Filters that do not introduce ringing in the time-domain response are also called linear-phase filters, because it turns out that having a phase change that is a linear function of frequency is necessary for this to be true. For most purposes of filtering biological signals a 4-pole or 8-pole Bessel filter is the way to go.

Passive, Active, and Digital Filters

Combinations of resistors, capacitors, and inductors are analog, *passive* filters. The simple RC filter is an example of this. For lower frequencies (<100 Khz or so) analog *active* filters are popular. These use op-amps and capacitors in clever feedback arrangements and have many useful properties, although as active devices they require a separate source of power to operate. Active filter design and construction is not for amateurs, and if you need one you can either buy general purpose filters installed into nice cases with clearly labeled controls etc., or pre-tuned modules that require only the addition of a power supply (as of this writing the company "Frequency Devices" is a good source).

As might be expected, digital filters are becoming increasingly common. Indeed, once a signal has been digitized there is no limit to what sort of processing a signal might be subjected to – filtering, processing, computation, it all blurs together. "DSP", or Digital Signal Processing, is a current buzz-word. In the simplest case an analog signal could be digitized, subject to a digital filtering operation, and then converted back into an analog signal (although this latter step is increasingly optional, as the digital filter output could be fed directly to additional computational or display processes within the same or to a different digital computer system).

The big "gotcha" with digital filters is time delay. All filters add some amount of delay to the signal they process. For analog filters this delay is typically a fraction of a period of a sine wave, and for most purposes can be ignored. However, for digital filters the delays can be substantial and not predictable from basic principles – they vary with the filter design and must be determined for each filter separately. For example, consider a digital filter that digitizes data at a 1 kHz rate. You might imagine that the time delay imposed by such a filter would therefore be 1 msec, but you would (almost certainly) be wrong. The digital filter might only take 1 msec per processing stage, but it might have 40 separate stages that the signal has to go through to get to the end, for a total delay input to output of 40 msec.

This delay, which is endemic not just in digital filters but in any device that uses digital techniques, can cause trouble in controlling rapidly changing processes. When recording data you must determine the delay for each filter and piece of equipment in turn and correct for this in the final analysis and display of data. Remember that the time delays of digital filters and other digital devices can change as a function of their parameter settings, so be careful! For some stupid reason the delays are not commonly found in equipment manuals, and are often hard to get out of the product support person you called on the phone. And even if they tell you the delay you should not trust them, but measure the delays yourself! Do you really want to retract five years worth of publications because the company salesperson was off by 20 msec in his informal chat with you on the phone?

Adding Multiple Sine Waves

One of the most widely misunderstood factors in frequency analysis is what happens when you add together a large number of sine waves of the same frequency but random phases. Popular conceit has it that as you increase the number of sine waves the result of adding them together becomes smaller and smaller, the idea being that sine waves of different phases will tend to cancel each other out. So when (for example) recording brain activity, an increase in a particular frequency of neuronal activity is considered de-facto proof that there is an increase in the underlying amount of neuronal synchrony.

Utterly false! When you add together a bunch of sine-waves of random phases, the resultant sum is always a perfect sine wave of exactly the same frequency, and whose magnitude will tend to *increase* by approximately the square-root of the number of input sine waves!!!! (After all, 1000 crickets may not be 1000 times louder than a single cricket, but they are a lot louder than a single cricket – even unsynchronized they do not cancel out completely!!!). This phenomena is illustrated in figure 13-11. So if you measure an increase in the amount of power at (for example) 40 Hz, this does NOT by itself mean that there is any sort of synchronous neural activity at 40 Hz – it might (or might not!) be that there is more total unsynchronized activity!

Of course if you *divide* by the number of sine waves you will see a tendency to average out (see figure 13-11, right half). It is a standard noise-reduction technique to average data over multiple trials, which causes uncorrelated activity in the recorded signals to decline and the correlated activity will be emphasized. For example, the electrical evoked potential on the surface of the head due to a flash of light (Visual Evoked Potential) is very weak, much less than the strength of the electrical noise for a single trial. The noise can be dealt with by recording many (thousands) of separate trials and averaging the electrical potentials centered around the time that the visual stimulus was flashed. This will reinforce the visually-evoked part of the signal and decrease the relative strength of the noise. But averaging requires an explicit division! When the electrical activity of a bunch of neurons superimposes via the electrically conductive media of the brain, there is nothing in the physics that causes a division, hence there is no averaging out! This effect is important but tricky. *Be warned.*

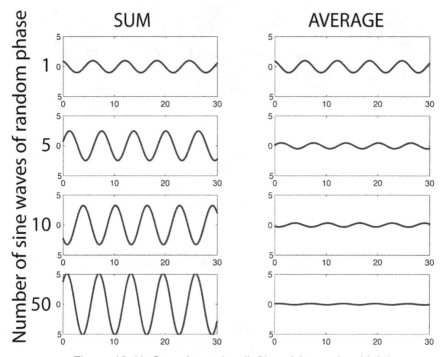

Figure 13-11. Superimposing (left) and Averaging (right)
Sine Waves of Random Phase.

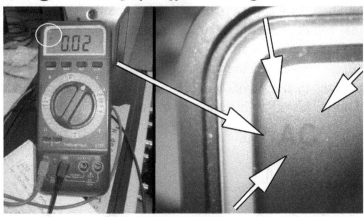

The multimeter is a really handy gadget to have around the lab. As it's name suggests, it can take the place of multiple different sorts of meters. Most can measure at least voltage, current, and resistance, and many can also measure capacitance or other things.

However, there is one thing that often bites newbies. Look at the picture on the upper left. See the teeny tiny "AC" in the upper left of the screen? It's blown up in the image on the upper right. It can be really hard to see, right? So here is the problem. Modern digital instruments are wonderful, but if you accidently hit a button you can set it into a mode and it can be hard – or often, impossible – to tell by looking at the instrument that it is no longer doing what you think that it is doing.

Normally you would want a multimeter set to "DC" mode. In "AC" mode it basically puts a capacitor in series with the input and only lets you know the AC value of the signal. Which is handy. Unless you didn't want that, and missed the tiny "AC".

Other things to remember about a multimeter: recall that voltage is measured ACROSS two points, and current is measured THROUGH a wire. To measure a voltage, switch the meter to voltage mode, make sure that the two test leads are connected to the voltage measuring connections, and put the two leads on the two points that you want to measure the voltage between. In this case, the meter acts like a very high resistance. To measure a current, switch the meter to current-measuring mode, make sure that the test leads are connected to the current measuring connections, break the connection that you are trying to measure the current in, and use the multimeter as a sort of substitute wire. In this case, the meter looks like a very low resistance wire.

Setting a meter to a current measuring mode and hooking the two terminals up to a high voltage difference is like shorting it out with a wire. The meter should be protected with a fuse but still, don't do that.

As electronic devices get more sub-menus and options the possibility of accidently putting them into a mode that does something weird increases. Most devices have a 'reset to factory default' option. When in doubt, use it.

Chapter 14. Circuit Models of Biological Membranes

A knowledge of electric circuit theory is valuable for a biologist not just to understand how electronic equipment works, but also to gain insight into the biophysics of processes with strong electrical effects. The flow of current through membranes is perhaps the most common example where electrical circuit models are vital for a complete understanding of what is going on.

Be warned, however, that electric circuit models can be misleading. In particular, electric circuits count only one kind of charge, while biological membranes have pores that are often selectively permeable to different ions. There are tricks that allow circuit models, which use only one kind of charge, to accurately model ion conduction through membranes, which use different classes of charges which cannot all flow through the same pores. But these models have their limits, and for some phenomena there is no alternative but to solve the differential equations governing ion flow directly.

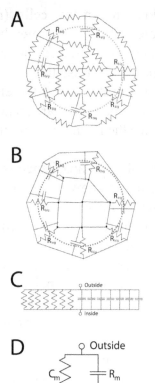

Perfectly Spherical Cells

The most common electrical model of a cell is the perfectly spherical cell. Real cells are lumpy complex structures, however, most of the resistance and capacitance is concentrated in the membrane (see figure 14-1). In other words, the electrical resistance across the cell membrane is much greater than the resistance between any two points within the cell, or between any two points outside of the cell. Therefore, it is approximately true that we can regard the voltage at all points inside a cell (or all points outside a cell) as being at roughly the same potential. As we recall from chapter 5, we can connect all circuit nodes that are at the same electrical potential with wires and not change anything. This makes the entire inside of the cell a single node, and the entire outside of the cell another node (see figure 14-1, panels B and C).

Figure 14-1. Steps in creating a circuit model of a (more-or-less) spherical cell.

Now all the membrane capacitance, and all the membrane resistance, can be treated as being in parallel. So the equivalent circuit model is just a single resistor in parallel with a single capacitor! For many purposes this approximation works quite well.

The time constant of the cell is

$$\tau = R_m \times C_m$$

This is the time constant that you would use for determining the time course of the cell responding to current injected by either a piece of membrane activated by a synapse, or by current injected through an intracellular electrode.

Note that many cells (most neurons, muscle fibers) generate action potentials when the membrane voltage goes over a set threshold. Until this threshold is reached the electrical properties of the cell are mostly linear, and you can use the model in figure 14-1. This linear range of operation is variously termed the electrotonic, graded, or passive operating regime. Action potentials are an all-or-nothing event, and hence nonlinear (because twice the input does not give you twice the output).

Membrane capacitance is usually homogenous over the surface of a cell, so if you know the capacitance in $\mu F/\mu m^2$ just multiply this by the surface area of the cell and you are done. Because conductances in parallel add, it is usual to work in conductances instead of units of resistance when dealing with a cell membrane. You can calculate a conductance/μm^2 as with capacitance. Alternatively, you can determine the total conductance per open ion channel, and multiply this number by the total number of open ion channels, which is the total number of ion channels times the probability of any given channel being open.

This analysis does a great job of estimating R_m and C_m for a more-or-less round cell. However, for a living cell there is active transport of Na^+ out of the cell and K^+ into the cell by the sodium-potassium pump. Combined with the differential permeability of the cell membrane to different species of ions, this results in an actively maintained cell membrane voltage. Circuit models cannot directly model the dynamics of ion movement through differentially permeable ion channels, but by using a separate conductance and battery for each species of ion a good functional model can be made (see Figure 14-2). Note that the "batteries" are the Nernst potentials for each ion. Also, we only need a single capacitor to represent the cell's capacitance because, unlike ion channels, the capacitance of a patch of membrane is not selective for different ions.

The resistance of pure lipid membrane is much greater than that of the ion channels, so it can be left out of the circuit model.

It is intuitive that the overall membrane voltage should be closest to the Nernst potential voltage of the ion channels that have the highest permeability, which accords with the biophysical Goldmann equation.

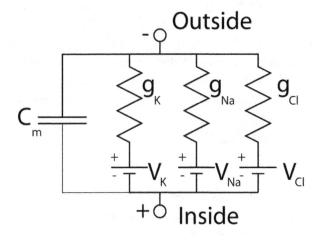

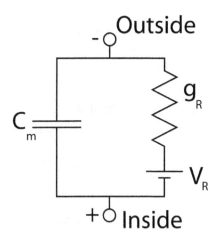

Figure 14-2. Equivalent circuit models of a cell that takes into account resting membrane potential.

Looking at the top half of figure 14-2 we remember that we can always replace a network of voltage sources and resistors with a single resistor and a single voltage source, as shown in the bottom of figure 14-2.

$$g_R = g_{Cl} + g_K + g_{Na}$$

$$V_R = (g_{Cl}V_{Cl} + g_K V_K + g_{Na}V_{Na})/g_R$$

By convention the reference polarity for the membrane potential is that the positive terminal is inside and the negative terminal outside. Thus, positive charge flowing into the cell makes it more positive, and positive charge flowing out of the cell makes it more negative. Note also that reference polarities for the "batteries" modeling the ion channels are not always consistently drawn: it depends on the text. As drawn in figure 14-2, all the virtual voltage sources are oriented with the positive terminal facing out. Thus, to match normal cell biophysics, V_K and V_{Cl} will be positive, which therefore tends to make the cell membrane potential negative, and V_{Na} is negative, which tends to make the inside of the cell positive. If you switched the reference direction of the Na "battery", V_{Na} could then be positive, but you would have to change the sign in the equation.

As ion channels open and close the corresponding lumped ion channel conductances will also change, and the model in figure 14-2 is moderately accurate at predicting these effects on changes in trans-membrane voltage. Note especially that if the membrane voltage is at the Nernst potential for its associated ion, that using Kirchoff's voltage law the voltage across the membrane subtracts away from the voltage across the ion channel, so the voltage across the associated conductance (resistance) is zero, and thus by $I = V/R$ no current will flow through that class of ion-specific channels, which is in accord with standard biophysics.

One of the more confusing issues in circuit models concerns the operation of the sodium-potassium pump. This "pump" uses energy to co-transport K+ into the cell and Na+ out of the cell. The pump is the engine that maintains the transmembrane potential, and if the pump is stopped the cell membrane potential will (eventually) decay to zero.

The NaK pump is often modeled as shown in figure 14-3, as a pair of current sources in parallel. The current source I_K represents the total sum of all active pumping of K+ into the cell, and the current source I_{Na} represents the total sum of all active pumping of Na+ out of the cell.

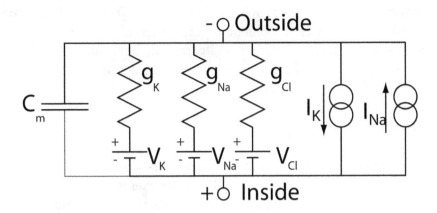

Figure 14-3. Circuit model of a membrane that incorporates the sodium-potassium (NaK) pump. This model can be misleading (see text).

However, if $I_K = I_{Na}$, by Kirchoff's current law the two current sources will cancel each other out, and will result in no net current being supplied to the rest of the circuit. Yet, the batteries will still create a net voltage across the membrane. What gives?

The answer is this: the NaK pump creates the transmembrane voltage by generating gradients of Na and K, and by the effects of differential membrane permeability to different ion species. This effect occurs even if the amounts of Na and K moved across the membrane by the active NaK pump are equal and opposite. This is the main force behind the creation of the trans-membrane voltage, and it is implicitly modeled in figure 14-2.

However, if there is an imbalance between the amount of Na and K pumped, then there is indeed a net current across the membrane due to the direct operation of the pump. The NaK pump ejects three Na+ for every two K+ imported, thus, it tends to make the cell membrane slightly more negative than it would if the pump were *electroneutral*. We say that the NaK pump is *electronegative*, but this is a second-order effect. The current sources on the right of figure 14-3 can predict the effects of current imbalances in the NaK pump, because it is only imbalances between these two currents that do not cancel out and can affect the operation of the rest of the model circuit. The overall function of the sodium-potassium pump is implicitly modeled in the "batteries" for the specific ions, and has nothing to do with the current sources in the circuit diagram.

In summary: circuit models of biological membranes are valuable, but must be interpreted with caution. Standard electric circuit models only use a single "flavor" of electrical charge, and cannot directly model biophysical systems that have different species of charge carriers that cannot all flow equally through the same paths.

Infinitely Long Cells

The above analysis dealt with cells that are more-or-less round. However, many cells or parts of cells are long and thin, for example, axons, dendrites, muscle fibers, etc. In this case the resistance from one end to another though the inside of the cell can be large relative to the resistance from the inside to the outside. Therefore you can't assume that all points inside a cell are at the same electrical potential, and replace it with a single node. You have to explicitly model the distribution of voltage as a function of position inside the cell. Real cells can have complex geometries, and are never infinitely long, but fortunately for many purposes assuming an infinitely long cylinder is a good working approximation.

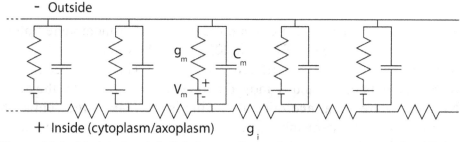

Figure 14-4. Model of an infinitely long cell or cellular process (axon, dendrite).

As shown in figure 14-4, the most common model is a one-dimensional finite-element model. In actual long cells/axons/dendrites, current and voltage vary continuously across the length. However, we can "lump" all the properties of small patches of membranes into discrete resistors, capacitors and voltage sources. This model is quite accurate as long as our subdivision into discrete segments is fine enough.

Each segment has a lumped resistor modeling the sum total of all ion channels, g_m. (Of course, a resistor and a conductor are the same thing, and use the same symbol, it's just how you measure the unit that changes. $g = 1/R$). There is also a lumped capacitor, C_m, and a voltage source, V_m, which models the effects of all the ion channels and ion concentration gradients.

84

The internal resistance of a length of the cell/axon/dendrite is g_i. Remember, if the total internal resistance from one end of the cell to the other was much less than the total resistance across the cell, the analysis would simplify to that shown in figure 14-1. Note also that the longitudinal resistance outside the cell is taken to be zero, i.e., a solid wire. That's because the conductive paths outside the cell are so many and have such a large volume, that the total resistance between any two points outside is generally very low, and so can be ignored relative to the other resistances. However this would probably not be the case for an isolated axon floating in insulating mineral oil.

Care with reference directions: a typical cell will have a resting membrane potential of around −70 mv. As drawn in figure 14-4, you would have to set V_m = +70 mv to model this. If you flipped the polarity of the voltage source then you could get by with V_m =   −70 mv. It's easy to draw it wrong. This doesn't matter if you know a priori what polarity the transmembrane voltage should be, but you can get in trouble when setting up formal equations.

Static Cable Equations, Steady State

The circuit model shown in figure 14.4 can be solved using standard circuit-theoretical techniques (use a computerized modeling program if you have several hundred segments!), or you can wire together a bunch of discrete capacitors, batteries and resistors and make a physical model. You will likely need to scale the values of capacitance and resistance to be different from those of a real cell, but the pattern of response will be identical. Initially we consider only the steady-state, where we can ignore the effects of the capacitors (replace capacitors with open circuits).

The model shown in figure 14-4 is linear. This means that you can use superposition. If you want to model the effects of injecting current through two microelectrodes, you need only model the effect of each electrode separately, and the result of both electrodes at the same time will be the sum of the effect of each separately. Also, you can model the effect of stimulating the cell separately from the resting membrane potential. In other words, if at rest the membrane is a uniform −70 mv, you can model the effect of injecting current to a membrane whose potential is initially a uniform 0 mv, then add in −70mv everywhere to get the total result.

It turns out that for an infinitely long cylinder there is a simple analytic solution. You can use this interchangeably with the finite-element model in figure 14-4: just sample the results of the continuous case at fixed intervals.

If you inject current into a long cell at a single point (as with either a microelectrode or an excitatory synapse) the voltage at that point will become more positive, and this effect will decay exponentially away from the point of excitation equally in both directions (see figure 14-5). The distance at which the voltage has declined to a value of 1/e times the peak value is called the *length constant*. The math is identical to the time constant discussed previously. In particular, after one has traveled more than a few length constants away from an electrical input, the effect will have decayed to nearly zero. This is why passive electrical signaling cannot be used for long distances, only the self-propagating nonlinear phenomena of the action potential can do that.

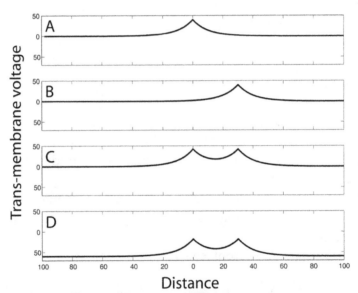

Figure 14-5. Pattern of response of injecting current at different points in an elongated cell or cellular process. A. Effects of current injected at one point. B. Current injected at a different point. C. Current injected at two points at the same time. D. Effects superimposed on resting membrane voltage.

As a linear phenomena, the passive electrical spread obeys superposition. Thus, the pattern of voltage along the length of an axon or dendrite in response to two spatially separated inputs is the same pattern as the sum of the responses to each separately. Additionally, you can analyze the effects of injecting current separately from the resting membrane potential, and add in a constant voltage along the entire length of the process to get the true potential distribution (see figure 14-5D). This is one of the advantages of analyzing a linear system.

The analytical expression for how voltage decays with distance in a long cable was first derived for underwater electrical cables, basically you just get an exponential decay with distance the same way that you get an exponential decay with time for an RC circuit. The "length constant" is defined by:

$$\lambda = sqr(r_m/r_a)$$

where λ is the distance at which the effect of injecting current has decayed to (about 37%) of it's peak value at the point where current is injected, r_m is the membrane resistance per unit length of cylinder in units of Ω-cm, and r_a is the axial resistance of a unit length along the inside of the axon, in units of Ω/cm.

So increasing the resistance of the membrane and decreasing the resistance of the cytoplasm increases the length constant e.g. causes electrical disturbances to travel farther before decaying. More than a few length constants away from the point of current injection and the response has decayed to negligible levels, which is why only action potentials can transmit information long distances in the nervous system.

Increasing the diameter of an axon causes membrane resistance to go down by the increase in the diameter, but the cytoplasmic conductance goes up as the square of diameter. Hence, all other things being equal, larger axons will have a longer length constant than smaller ones. While we are not directly discussing action potential propagation here, axons with longer length constants will also have faster action potential propagation because a given patch of depolarized membrane will be able to depolarize other patches a greater distance away, hence increasing the speed at which an action potential can propagate.

Shown above is a picture of a potentiometer. You can fit a knob onto the metal shaft to make it easier to twiddle. This is the real physical device whose circuit diagram is shown in Figure 5-5. It has three leads. The middle lead is the "wiper", corresponding to terminal "c" in Figure 5-5. Potentiometers are built into nearly anything, you likely won't need to mess with one, but there are cases where they are handy. If you connect one outside terminal to ground, and the other outside terminal to a positive voltage, such as +9 Volts, then the middle terminal will have a voltage that you can continuously vary between 0 and +9 volts just by turning the shaft.

There is not much to go wrong here, but there are some issues. First, potentiometers come in two flavors: linear and logarithmic ("log"). The log ones are useful for things like volume controls in stereos, because our hearing is logarithmic. But if you don't know this you can be surprised. Also, potentiometers ("pots" for short) are mechanical and can wear out. Old pots that have been lying around in someone's bottom drawer since the Grover Cleveland administration are particularly suspect.

Sometimes there will be small squarish or rectangular pots built into the guts of an electronic circuit, and designed to be moved not with a knob but with a small jewelers screwdriver. These are called "trim pots", and are usually designed to be changed only rarely to fine-tune some aspect of an instrument. Don't mess with anything like this unless you have read the manual first…

Chapter 15. Electrodes

Electrodes: where circuits meet tissue. In circuits electrical current is carried mostly via electrons flowing in metallic wires, and in tissues current is carried mostly via electrically charged atoms and molecules of various sorts in aqueous solutions. Connecting the two requires a point of interface whose properties can be complex.

Electrode design is most critical when you need to make accurate DC measurements, such as of the resting membrane potential of a cell. If you only care about the AC part of a signal, for example detecting whether or not an action potential has occurred, then the issues of tissue-electrode interface become less critical (though the wrong electrode can still ruin your day). Electrode design is also less important for high voltages. If you just want to shock a peripheral nerve with a pulse of several volts you could use almost anything for the electrode. Make a careful measurement of DC potential to the millivolt and you need more care.

Metal Microelectrodes

A variety of metals are used for making microelectrodes. Typical choices include tungsten, stainless-steel, platinum, and platinum-iridium. Sometimes the tips are plated with gold. In general the electrochemistry at the metal-solution interface is complex, and capable of changing with time or with the direction and magnitude of current flow. Additionally, current passage typically causes the electrolytic release of gas bubbles, which both alters the electrode tip characteristics and can cause obvious problems with intracellular recording. These problems are most serious for DC or low-frequency signals, and much less so for transient or higher frequency signals. Thus, metal microelectrodes are most often used for the extracellular recording of transient events, such as the action potential of a neuron, where the precise DC level of voltage is not important.

The bottom line: as long as you only need to know that an event has occurred a metal microelectrode is fine. The minute that you need to know the precise magnitude of the signal to the level of a millivolt simple metal microelectrodes are unsatisfactory.

Most of the energy in the electrical activity of an action potential is in the frequency band of 300Hz to 5 Khz. Hence, when recording action potentials, you can use a filter that only passes this frequency band and that rejects the DC level, as well as the 60/50Hz power-line frequency. While intracellular action potentials span a range of many tens of millivolts, extracellular action potentials are typically only a few hundred microvolts (fraction of a millivolt) in amplitude.

Extracellular recordings of action potentials does have a down side: the extracellular waveform can vary with the precise location of the electrode tip relative to the cell body, and also with the rate of firing. Additionally, when recording the action potential of non-isolated neurons, the signals from multiple neurons can be superimposed onto the recording. This problem is typically, if imperfectly, handled by carefully moving the microelectrode tip to be close enough to one neuron that its action potential is routinely several times that of the action potential of the nearest neighbor. The general issue of reliably separating out the signals from multiple neurons from extracellular recordings is complex and there is no currently agreed-upon best technique. A good reference to start with would be Harris et al, J. Neurophysiology 84: 401-414, 2000.

The Silver-Chloride Electrode

Perhaps the most common interface of electronics to tissue where precise voltage levels are required is the silver-chloride electrode. Current in wires is carried by electrons, and current in biological solution is carried by chloride ions Cl- (see figure 15-1). Initially the current is carried by electrons in copper wire, but this is changed to silver wire. The end of the silver wire is

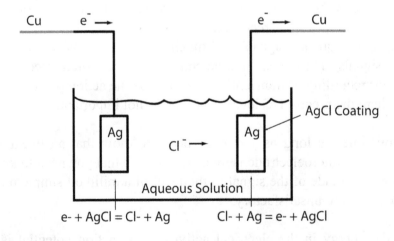

Figure 15-1. The Silver-Chloride Electrode.

coated with Silver Chloride AgCl, and the incoming electrons are converted to silver and free Chloride ions. The ions are free to move through the aqueous solution, and at the other electrode the process is reversed, chloride ions are converted to electrons and silver chloride.

There are no gas bubbles, and the apparent DC voltage at each interface is stable and relatively independent of current direction, i.e., these interfaces do not act as rectifiers.

Note that this only works for solutions that have Cl- in them (typically a minimum of 40 mM Chloride is required). Also, if the current passes one way for too long you can run out of AgCl at one electrode, exposing bare silver wire which can affect many biological processes and which results in a more unpredictable wire-solution interface. A reminder, in the diagram electrons and chloride ions are moving left to right, current is of course moving in the opposite direction.

The big problem is that at the interfaces between different materials you get electrical potentials, much as batteries generate a voltage by bringing electrochemically dissimilar materials together. These effects are usually modeled as a constant voltage source, see figure 15-2.

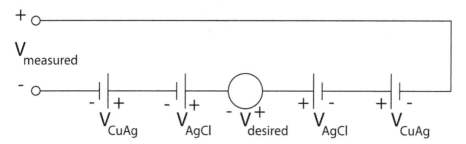

Figure 15-2. Electrical model of the situation in figure 15-1. In principle, the voltages due to the metal-metal and metal-electrolyte interfaces cancel out.

You are interested in $V_{desired}$, but you only have access to $V_{measured}$. However, if the voltages due to the copper-silver interfaces (V_{CuAg}) and the metal-electrolyte interfaces (V_{AgCl}) are equal and opposite, then these will cancel out and $V_{measured} = V_{desired}$.

The interfaces between the copper and silver wires will be equal and opposite under any realistic circumstance, so can be safely ignored as these will always cancel out (though if you put a match under one junction, you will unbalance the situation and effectively create a thermocouple!). However, the interface potentials at the silver-chloride electrode depend upon the concentrations and mobilities of the ions at the point of contact, the temperatures etc. If the concentration of chloride is different at one of the electrodes (for example one end is inside an electrolyte-filled microelectrode) then the voltages due to the two electrodes will not be equal and opposite, they will not cancel, and $V_{measured}$ will not equal $V_{desired}$.

If you are only interested in changes in electrical potential you can simply remove the net DC potential of the imbalance with an offset control. However, if the concentrations of ions changes in any part of the circuit, or if you need to know the absolute voltage, this approach is inadequate.

Ch. 16. Shielding and Noise

Life is never as simple as the diagrams in textbooks. As we have emphasized throughout, electrical circuit diagrams are simplified models of complex physical phenomena that, under the right conditions, can provide very accurate fits to reality – or at least, fits that are close enough.

But it does occur, more often than not, that the complexity of the real world gets in the way of what we would want to record or do. In principle we could create a very complex physical model that encompasses all that we understand about all aspects of a system and solve using a network of supercomputers. In practice we lump all aspects of an electrical signal that we don't want/didn't anticipate, call it "noise" (just as any plant that you don't want is a "weed"), and develop some rules-of-thumb for dealing with it.

Roughly, noise may be split into two categories. "Intrinsic Noise" is noise that is intrinsic to the electrical and thermodynamic properties of the system under study, for example, the fact that electric charge is not purely continuous but actually consists of a finite number of electrical charges can cause a jitter in electrical measurement. Intrinsic noise is usually anticipated and dealt with in the engineering of any commercially available piece of equipment, and hence is not generally something most biologists need to worry about. The main exception is when the biological process you are measuring is itself a prime source of intrinsic noise.

"Extrinsic Noise" comes from the interaction of the equipment/system with the rest of the world, for example, the 60Hz "hum" that is picked up by interactions with power lines (50Hz in Europe. Life would be much simpler if we did our experiments using only battery-powered equipment in the middle of the Gobi desert). A better term for extrinsic noise would perhaps be *interference*. Interference is omnipresent, and will prove to be a source of effort for even the most determinedly non-engineering minded biologist. Your equipment may be the best obtainable, designed by electrical engineers of protean genius and no social life, but you still have to hook this equipment up with wires to other equipment and to biological preparations – and with these wires comes the influences subtle and gross from the myriad of other things that exist in our world.

Signal-to-Noise Ratio/SNR

Noise is often quantified by the Signal-to-Noise Ratio, or SNR. If the signal is ten times the magnitude of the noise, then we would have an SNR of 10.

However, our old friend the decibel is often used here. As you recall, the decibel is a log scale measure of the ratio of two quantities, which makes it useful when there is a large range of ratios. There is often confusion about whether a decibel refers to a ratio of power or a ratio of signal magnitude. When in doubt assume signal magnitude.

SNR (Decibels) = $20 \log_{10}$ (Signal/Noise)

Hence, a signal-to-noise ratio of 10 is 20 dB, a ratio of 100 is 40 dB, etc.

Thermal Noise

Thermal noise (also known as Johnson Noise or Nyquist noise) is the noise that comes from the random movement of electrons (or other charge carriers). As you would expect this noise is reduced at lower temperatures – hence some ultra-sensitive electronics are sometimes cooled to cryogenic temperatures.

Thermal noise is a random fluctuation, and can thus be reduced by averaging over a sufficiently long interval, i.e., it has a DC value of zero. The RMS (Root-Mean-Square) value of this AC noise signal is given by

v_{RMS} = SQR(4kTRB)

Where v_{RMS} is the expected RMS voltage of the random thermal noise over a long interval, k is Boltzmann's constant (1.38×10^{-23} joule/°K), T is the temperature in degrees Kelvin, R is the resistance in ohms, and B is the bandwidth in Hz under consideration. ("SQR" is a shorthand for square-root).

Hence this source of noise is reduced at low temperatures, low resistance, and low bandwidth. The issue of bandwidth is a more general one: to increase the precision of a measurement it never hurts to narrow the focus of your instruments to what you are trying to measure. Bandwidth is the difference between the highest and the lowest frequency component of a signal, it is measured in units of cycles-per-second or Hertz.

94

Thermal noise is *white noise*, i.e., it has constant magnitude per Hz of frequency, and the value of the noise at one moment in time cannot be used to predict the value at a later time. In engineer-speak we would say that the random process has no *memory*.

A 1 MΩ resistor at room temperature and a frequency band of 0 to 10 KHz has thermal noise of 40 µV RMS. This is not much: but if present at the input stage of an amplifier that has a gain of 10,000 it comes out to 0.4 volts!

Moral of story: when amplifying very weak signals, the most critical stage of amplification is the first one. Make the pre-amplifier low noise and the rest is (relatively) easy.

Shot Noise

Shot noise comes from the statistical randomness of discrete charge carriers crossing a junction of some sort. It is most often defined for the semiconductor junctions found in transistors. The size of the effect is:

$$i_{shot} = SQR(2qIB)$$

Where i_{shot} is the rms value of the shot noise current, q is the charge on an electron (1.59 x 10^{-19} Coulombs), I is the DC current in amps, and B is the bandwidth of the signal. Like thermal noise shot noise is white noise. Unlike thermal noise, shot noise depends upon the DC current through a junction of some sort, and is lessened by lowering the current.

Flicker Noise

In contrast to thermal and shot noise, flicker noise is not white noise. It decreases with increasing frequency proportional to 1/f, hence its other name of "1/f noise". The sources of flicker noise are not well understood, so yet another synonym is *excess noise*. The 1/f nature of flicker noise means that, for many conditions, flicker noise is dominant at low frequencies. This noise often also includes the effects of slow drifts in circuit parameters. The properties of flicker noise mean that it is often best to take critical measurements in a frequency band away from DC, if possible.

Extrinsic Noise/Interference

The paths by which external interference may slither their way into your system are various and manifold. The world is complicated. We will barely scratch the surface of the topic here. Before covering some specific cases a few heuristics (rules of thumb) for avoiding interference:

-> Do not place large motors (centrifuges, fans, etc.) or transformers (power supplies, "wall-warts") near anything that makes sensitive measurements. The floor polisher out in the hall may have an effect too.

-> Use a single ground point for all equipment. Run separate wires to ground from each piece of equipment, i.e, do not "Daisy Chain" ground connections.

-> Try wrapping delicate equipment in aluminum foil which you hook to ground at a single point with a wire. Try placing the entire recording setup in a grounded screen-wire enclosure.

-> Be prepared to swap wires around: use trial-and-error, it often matters which piece of equipment is plugged into which socket, or which cable goes over and which goes under.

-> If at all possible set up your equipment so that all long cables are carrying large (on the order of volts) signals. Don't try to send a microvolt signal across the building! Don't hook up a patch-clamp electrode to the preamplifier with 20 feet of coaxial cable. Weak signals should have the shortest possible cables before they are amplified to larger signals appropriate for standard instrumentation.

-> Read The Fine Manual (RTFM).

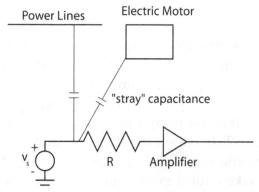

Figure 16-1. The problem of capacitive coupling.

Capacitive Coupling

This is the most common source of interference for a biologist. As you recall, a purpose-built capacitor is two metal plates separated by a small insulating gap. The value of the capacitor is maximized by making the plates large in area and small in separation. However, *any* two wires will have some capacitance between them. As wires are typically small in area and relatively far apart, the effective capacitance between two wires is similarly small, however, depending upon the values of the impedances in the different parts of the circuit and the frequency range of the signals, this capacitance can have an effect. You can model these effects as a small capacitor connecting two wires (see figure 16-1). We term this virtual capacitance a "stray", or sometimes a "parasitic", capacitance. For large signals and low values of resistance the effects are small, for low-amplitude signals in high-impedance circuits the effects can be very large.

Shielding

The simplest way to reduce capacitive coupling is to physically move the interference away from where you are trying to make delicate measurements. Also, you can *shield* a sensitive part of a circuit with a metal cage that is connected to ground (see figure 16-2). The capacitive coupling with the interference is thus to the shield, completely eliminating the coupling to the circuit of interest.

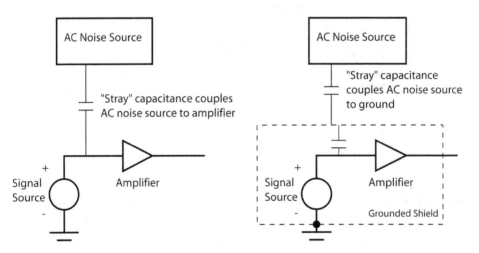

Figure 16-2. Principles of Shielding. Left: "Stray" capacitance couples a time-varying (AC) noise source to an amplifier. Right: a grounded shield shunts the currents induced by the noise source to ground. There is still a small fixed stray capacitance from the amplifier to ground, but as the ground is at a constant voltage, it does not add noise.

There is, of course, some capacitive coupling of your own circuit to the shield, but as the shield is connected to ground (which is at constant zero volts) this coupling will not add any time-varying noise. The shield can have large holes in it, it can be made of metal screen wire or mesh, the only critical thing is that the shield has to completely enclose the circuit to be shielded, it must be continuously conductive i.e. it cannot consist of metal sides connected with non-conducting plastic tape, and the shield must be connected to ground. Such a shield is sometimes called a *Faraday Cage*.

Ground Loops

While simple in theory, in practice shielding can be tricky. The biggest problem in shielding (other than forgetting to ground the shield, or not noticing that the connection of the ground to the shield has come loose!) is the dreaded *ground loop*.

Correct: Shield Grounded at a Single Point

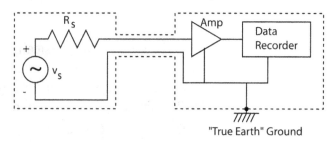

WRONG: Shield Grounded at Two Points

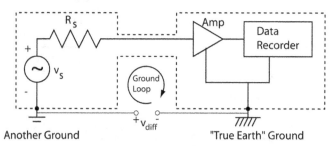

Figure 16-3. Ground Loops.

Consider the circuit in figure 16-3 which illustrates a voltage signal source v_s with source resistance R_s, connected by a cable to an amplifier and a data recording system (such as a computer or a meter). The shield covers everything and is only grounded at one point, here termed "True Earth" ground but it is not really necessary that this actually be a wire going into the dirt: the main thing is that the ground is at a single point.

In the bottom half of the figure we see an incorrect grounding. Both the ground and the shield are grounded at two separate points. This leads to several problems. First, the two grounds may be at different potentials, leading to an offset voltage v_{diff}. The difference in voltage between the two grounds can result in a spurious signal being recorded. In addition, the gap between the shield and the external connection between the two grounds (which may be through the ground wires running through the walls of your lab) can act as a large antenna, and any time-varying magnetic field passing through this gap (loop) can induce time-varying currents through the wires, which as even thick copper wires have some finite resistance, will result in time-varying voltages injected into the circuit as noise.

The proper geometry and connections of shields and grounds can be very confusing. Be prepared to use a lot of trial-and-error.

Signal Guarding

For very low-level signals sometimes shielding alone is not enough, and you have to resort to using a *guard* in addition to a shield. An elegant technique, but there are stability issues that mean that, unlike shielding, you will probably never design and build a guard yourself – it will more likely be included as part of a sophisticated measuring system that you will buy - but it is still important to understand what it is.

The key idea is similar to that used in symmetry. When two nodes of a circuit are guaranteed to always be at the same voltage, then regardless of whether the two nodes are connected by a resistor or a capacitor, or both, no current will ever flow between the two nodes. In this case, it is as if the resistor or capacitor isn't there!

If we enclose the input stage of a sensitive amplifier in a metal tube, and connect the output of a unity-gain amplifier to the tube, we can force the metal tube to always be at the same voltage as the input (see figure 16-4). Thus, there will be no current flow between the tube and the signal wire due to either stray capacitance or resistive leaks. We have in effect cancelled out the stray capacitance! Be warned though: improper construction or connection of a guard can cause feedback oscillations (sort of like when a person talks into the microphone of a public address system and causes a "howl") so they are trickier than mere shields. Obviously, you can put a shield around the guard for good measure.

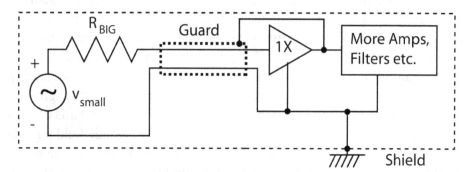

Figure 16-4. Using a signal guard with a small-voltage, high-impedance signal source. The output of the unity-gain pre-amplifier drives later stages of amplification but also connects back to a guard surrounding the inputs, thus forcing the guard to be at the same potential as the input and eliminating any resistance or capacitive effects on the input wires.

Differential Amplifiers

Differential amplifiers – often also called instrumentation amplifiers – are a key part of any sensitive measuring electronics. Recent improvements in differential amplifier design have to some extent limited (though not eliminated!) the need to use shielded rooms. We have already encountered differential amplifiers in the guise of op-amps. Differential amplifiers designed for instrumentation techniques have a lower and much more accurately controlled gain than is the case for a standard op-amp.

The key concept is this: external interference will often cause the same noise to be added to both the signal and the ground wires of a piece of equipment. By using an amplifier that only responds to the difference in the voltage between two points, we can amplify only the signal and ignore the noise. As illustrated in figure 16-5, a voltage signal has external noise added to it. However, the external noise also adds to the ground as well as the + terminal of the source.

Amplifying only the difference between the two extracts the original signal and ignores the noise that added equally to both wires. In this case the noise and the signal were shown occurring at different frequencies for clarity. However, unlike filtering, a differential amplifier will reject the noise (common-mode) signal even if it were at the same frequency as the desired (differential mode) signal.

A differential amplifier is designed to ignore a common-mode signal, i.e., to not respond when both inputs do the same thing. It is designed to respond only to the difference in the signals. All real differential amplifiers do respond to some degree to a common mode voltage, but only very weakly. One measure of the quality of a differential amplifier is the Common Mode Rejection Ratio, or CMRR. This is the ratio between the sensitivity to a differential signal versus the sensitivity to a common-mode signal. A differential amplifier with a CMRR of 120 dB will give the same response to a differential-mode signal of 1 microvolt as it will to a common-mode signal of 1 volt. The quality of instrumentation amplifiers is also measured by its input impedance, which should be as high as possible to avoid influencing the system being measured.

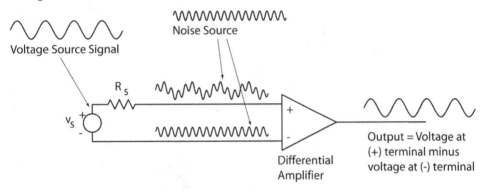

Figure 16-5. Example of differential amplifier operation.

Differential amplifiers are not always labeled as such, but they are very common in the first stage of amplification of many systems such as microelectrode preamplifiers, EEG amplifiers, etc. Many systems for inputting analog voltages into computer systems can be configured to run either "single ended", i.e., as conventional amplifiers, or "double ended", i.e., as differential amplifiers. Typically running a computer Analog-to-Digital converter in differential input mode halves the number of input channels, as you need two separate inputs per channel rather than just one input and a single ground shared between all inputs.

Be careful not to confuse differential amplifiers with Op-amps. An Op-amp is indeed a differential amplifier, but with very high (X100,000 or more) and poorly controlled gain: they are almost always used with feedback to limit the overall circuit gain. A purpose-built differential amplifier has a limited, stable and precisely calibrated gain (like X0.1, X1, or X10) and is usually employed purely as an amplifier. The symbols for the two can be misleading sometimes.

Magnetic Shielding

Electricity and magnetism go hand-in-hand. A changing electric field produces a changing magnetic field and vice-versa, and an electrical current always produces a magnetic field. Radio-frequency interference is a propagating interaction of electric and magnetic fields. However, because of the properties of commonly used materials, and because there is no such thing as magnetic "charge", there are differences between shielding from mostly electric interference and mostly magnetic interference.

In particular, magnetic fields are very hard to shield against. Thin metal foil or screen wire is great for shutting out the effects of stray capacitance, but does not shield against the effects of magnetic fields. The best way to deal with magnetic fields (such as are generated by electric motors, transformers in power supplies, or CRT monitors) is to remove the source of interference as far from the location of any delicate measurement as possible. Using battery-powered equipment near where you are making sensitive measurements can be a good idea. If you have to, you can buy a shielding material called "mu-metal" that provides some magnetic shielding, but is a lot more expensive, a lot heavier, and a lot less effective at shielding from magnetic fields than aluminum foil is at shielding out electric fields.

Using filtering and differential amplification works as well for removing magnetically-induced interference as it does for any other kind of interference. In addition, magnetic interference can be greatly reduced by keeping all wires close together. This is because wires in a circuit that are spread far apart act as a loop that "catches" a large amount of magnetic field, like a big antenna. Holding two wires together minimizes the area that can gather magnetic field. Even better, using two wires that twist around each other ("twisted pair") results in equal and opposite magnetically-induced currents with each twist, which tends to cancel out the interference.

Chapter 17. Computerized Data Acquisition

We have already touched a bit on computerized data acquisitions in chapter 12. However, this topic is important – and it becomes more so every year – therefore in this chapter we review and expand on this topic.

Computers are wonderful devices that have revolutionized the use of electronics in science. But with this power comes a host of new problems and issues that must be dealt with. If you do not understand the fundamental principles of a computerized data acquisition system, it is all too easy to make serious mistakes that will come back later and bite you. This chapter does not teach you how to program or design computer-based data acquisition and experiment control systems, but instead goes over what you need to know to avoid making mistakes in hooking them up and using them.

Continuous vs. Discrete Signals

Real-world signals are essentially continuous both in time and in amplitude (There is the concept of a shortest possible time and a shortest possible size – the Planck time and the Planck length – but these are so small that biologists can generally ignore them). However, computers cannot deal with continuous numbers, they can only handle real-world signals if they make them *discrete*, that is, force each signal to a limited number of possible states. You can think of the real world as a smoothly-sloping ramp, where you can stand anywhere, and the computer as a staircase were you can only stand on one step or the other, and half-steps are not allowed (see Figure 17-1).

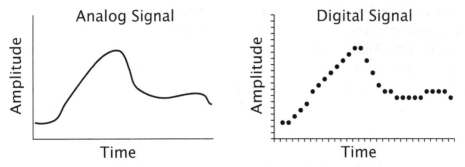

Figure 17-1. Analog (real-world) signals vary continuously in both time and amplitude. Digital signals are only defined at fixed points in both time and amplitude. The digital signal has a fixed number of bins in time and amplitude, and the signal cannot lie between two adjacent bins: it must be completely in one or the other.

The fundamental reason for this is that a truly continuous signal carries infinite information. Consider making a mark on the length of a toothpick. Suppose that you measure the distance of this mark from the end, and that it is 1.43223 centimeters. You could take these numbers and make them into a code. If you used pairs of numbers to code for a letter, the measurement at the precision above could represent three letters. This is not very much information. But if you had infinite precision, you could code for an infinite number of letters. Which cannot fit into a digital computer, as these have only finite numbers of binary bits.

There are less fundamental but more practical reasons for why computers can only deal with raw data in a discrete format, such as the limitations of the circuitry that converts an analog into a digital signal, but the bottom line remains that when you convert a real-world analog signal into a digital signal, you must quantize it to discrete levels.

The computer itself does not understand time or voltage. It orders the data in time by an index, and creates a single binary number for the magnitude at each sample point. You need to assign what these mean in physical units after the fact (though many purpose-built systems do this for you). Consider the following data:

1: 0
2: 37
3: 23
4: 55
5: 4

There are five data points. These points could represent signals that were sampled every second, or every hour, or every millisecond. Or every 0.23 milliseconds. And the value could have the units of volts, or microvolts, or milliamperes, or 0.03 of a milliampere. Some programs will automatically handle the scaling from the dimensionless signals that computers use to the corresponding units in the real world, but it is useful to remember what the computer is actually working with.

Discrete amplitude: Analog-to-Digital ("A/D") converters are specified in bits. A 12-bit A/D converter takes a continuous voltage and quantizes it into one of $2^{12} = 4096$ discrete steps. The range over which an A/D converter can accept data will vary/can be varied. If a 12-bit A/D converter is specified to accept inputs over a range from zero to +10 volts, then a zero voltage will give you the number 0, and +10 volts will give you the number 4095 (if you start at 0, then with 4096 steps the maximum number is one less than the number of steps). Each step will be 10/4096 = about 2.44 millivolts. So voltage changes on the order of a millivolt will not be reliably recorded.

If a 12-bit A/D converter takes input from -10 to +10 volts, then each step will be about 4.88 millivolts. Depending on how the converter is configured, -10 volts could be 0 and +10 could be 4095, or -10 volts could be -2048 and +10 volts could be +2047. Voltages over +10 or less than -10 will be out of range: this will likely lead to clipping, i.e., input voltages greater than +10 volts are recorded as +10 volts, and input voltage less than -10 volts will be recorded as exactly -10 volts. Hence, using an A/D converter requires trading off the range of input vs. the resolution. Making this trade-off and setting the correct range and precision of your A/D converter is an absolutely critical part of using computerized data acquisition for biomedical applications,

Depending on how precise your measurements need to be, you might need to specify the sensitivity of your system in "microvolts per bit" or some such. For example, the literature on electroencephalography typically requires that you specify that your system had a sensitivity of (for example) "0.1 microvolts per bit". In this case we are only referring to the least-significant bit of the digital number: if the signal at the electrode increases by 0.1 microvolts, then the number as recorded by the computer (AFTER it has passed all stages of amplification etc) will increase by the integer value one.

Some systems will convert the native binary number which is the output of an A/D converter into a real floating point number or convert to some appropriate units (gallons, degrees Celsius, pH, etc) but the native unit is binary.

More bits gives you more resolution at the same range, but usually at the cost of either more money, or lower speed, or both. Still, as technology advances, faster and higher-resolution A/D converters keep getting developed.

Floating Point Numbers and Symbolic Math

The above discussion only applies to the conversion process from analog to digital. This conversion process limits what aspects of a real-world signal can be imported into a computer. If the real world signal is 1.2 millivolts, and the computer records this as 1, you can't get back the 0.2 difference by using more precise numbers after the fact. However, once analog data has been converted into a digital format, further processing can be done with greater precision. In particular, the internal processing by a computer can use <u>floating-point numbers.</u> A typical double-precision floating-point number on a modern computer has a precision of one part in 2^{53}. This is far more precise than any existing or likely A/D converter. The issues of how the precision of using floating-point numbers affects computational accuracy is a topic beyond the scope of this text, and if interested the reader should consult a source on <u>numerical analysis</u>.

A number like pi cannot be represented with complete accuracy using a standard computer floating-point number. Pi is 3.14156.... to an infinite number of digits, and you can't represent an infinite number of digits on any finite machine. However, pi can be represented symbolically, just as a human mathematician uses the symbol "pi". These programs are said to use <u>symbolic math</u>. In general symbolic math programs are more suited to solving proofs or doing theoretical physics than they are in dealing with real-world data, but it is useful to be aware of the difference.

Guard Bits

A real world analog-to-digital converter is a device that, like all real devices, has many limitations and imperfections. One limitation is linearity: the voltage difference between one step and the next might not be constant across the entire range of input values. Read the fine manual (RTFM) that came with the A/D converter if you want a full list of the different kinds of errors that can afflict an A/D converter.

However, one of the most prevalent limitations of an A/D converter is that the least significant bit (or two) is unreliable. Imagine that you shorted out the inputs to an A/D converter: that is, you connected the plus input directly to ground. In this case the input voltage should be zero, and the binary word created by the A/D converter should also be zero. But if the least significant bits are unreliable, a plot of recorded voltage vs. time will show the voltage bouncing around in an apparently random manner (see Figure 17-2).

What this means is that you should give yourself some slack. If an A/D converter is set up so that the difference between steps is 1 millivolt, you should not depend on it measuring more accurately than (approximately)

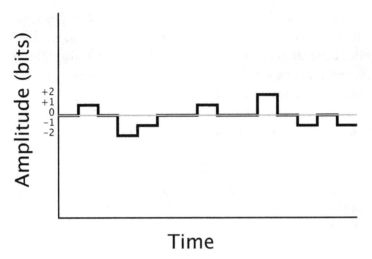

Figure 17-2. Even though the input voltage to an A/D converter is zero volts (gray line) the output often randomly drifts around this value by a bit or two. Thus, the least significant bit (or two) of an A/D converter is often unreliable.

4 millivolts or so. If you need to take measurements with a precision of a millivolt it would be good practice to either increase the amplification, or get an A/D converter with a higher number of bits, so that you have (something like) 0.2 or 0.3 millivolts per bit.

Having more bits of accuracy than you theoretically need is called having "guard bits". In essence, you need to assume that your system will not work perfectly and build in tolerance for small errors.

Sampling rate/aliasing

If the input to an A/D converter has frequency components that are high relative to the sampling rate, you can get aliasing (See again Figure 12-2). Aliasing can take many forms and surprise you if you don't know what to look for. The Nyquist sampling theory tells us that we need to sample at least twice as fast as the highest frequency component. Thus, if the highest frequency component in your signal is 100 Hz, in principle you need to sample at a rate of 200 Hz to avoid aliasing.

However, this is a theoretical figure that in practice is not fast enough. You want to sample faster than this Nyquist frequency. So: if you are sampling a signal that has a maximum frequency of 100 Hz, you might want to sample at 1000 Hz, and have an anti-aliasing filter with a cutoff frequency fc=200Hz, to give yourself some cushion. Don't try and push the limits. Remember also: just because the cutoff frequency is 200 Hz, it will still pass some amount of the frequencies above this. Remember also that, for most purposes, the optimal class of low-pass filter for this purpose is the Bessel, or linear phase, kind (see again Chapter 13).

Analog ("low level") and digital grounds

Reference has been made to the difficulties involved in arranging ground connections correctly. Grounding is perhaps the single biggest source of difficulties in the use of electronics in biology. This is complex and often there is nothing for it but to use trial and error to figure it out.

However, with A/D systems there is a further complication to worry about. The standard ground on a digital computer system is often corrupted with noise from the internal operation of the computer. Thus, many A/D systems have both a regular ("digital") ground and also a separate ground that is variously termed the "Analog ground" or "low-level ground". Exactly how this ground relates to the regular ground varies from system to system. The bottom line: if you use the regular ground on an A/D converter, and neglect to hook up the analog ground, it is possible that your system will act as if the ground is not hooked up at all, which can either give you no signal, or (even worse) a corrupted signal due to capacitive coupling.

Moral of the story: read the fine manual. See if there is anything labeled "low-level ground" or "analog ground". "LLGND" or "ANAGND" are also possible notations that should attract your attention.

Conversion delay

Nothing works instantaneously. Analog instrumentation – filters, amplifiers, etc. – often have processing delays that are so short that, for many biological processes, they can be ignored (although you should always check). However, digital devices typically have longer delays. What's more, these delays are often much longer than the sampling time would suggest.

Consider a pipeline that is transporting oil at a rate of a barrel per second, and that contains a total of 1000 barrels. Every second a new barrel comes out one end, but the total amount of time that it takes any one barrel to travel the entire length of the pipeline is 1,000 seconds. This sort of <u>pipelined processing</u> is routinely used in digital signal processing systems. (You can also think of it as an assembly line). So a system that acquires data at a rate of one sample every millisecond might have a total delay of (for example) four milliseconds. That is, when the analog voltage at the input changes the computer does not record this change until four milliseconds later.

For many applications the delay times are too short to be important. Also, if all of your data is recorded with the same fixed delay, then the relative timing of the signals is still good. But suppose that you had a digital bit turning on a stimulus with almost no delay, and a complex digital filter recording the response with a delay of 20 milliseconds. If you had not checked the delays ahead of time, you would report stimulus-response timings that are 20 milliseconds slower than they really are.

Here is another example: A system uses a video camera to track eye movement, outputting two analog voltages that indicate the horizontal and vertical eye position. The video system collects a new image every 16.7 milliseconds. However, because of internal pipelined processing, the delay between when the eye moves, and the analog output voltage reports this movement, is on average 32.4 milliseconds. This time delay can add significant error if you do not measure it and subtract it off.

ALWAYS CHECK SYSTEM TIMING WITH A REFERENCE STANDARD. For example, if you were using a digital bit to trigger a stimulus and then recording the response, set your system up to record the triggering pulse after it has gone through all the processing stages that your real biological signal would.

Channel Skew/Multiplexing

A typical A/D system available today will have multiple input channels, often a power of two, such as 4, 8, 16, or 32. For various technical reasons, these systems often (not always, but often) have only a single A/D converter and they create multiple inputs by multiplexing them into the single A/D converter. In other words, when getting input from multiple sources, the data acquisition system connects an input to the A/D converter, makes a conversion, then it hooks another channel up to the A/D converter, makes a conversion, and so on.

There are two potentially important consequences of this. One is that the maximum sampling speed will be slower when recording from many channels than from one. The other is that the time at which the channels are sampled might not be the same. This channel skew is usually small enough that it can be ignored, but not always, and you should be aware that it exists.

Single-ended vs. Differential

We have already talked about differential amplifiers in Chapter 16. Many A/D systems have the option of working in single-ended mode, where each input is referenced to a common ground, or in differential mode, where the input is the difference between two channels. Typically in differential mode you get half as many channels as in single-ended mode. The advantage of differential mode is that, at least for some conditions, you can have superior noise immunity (see again Chapter 16).

The big danger is that if you hook up an A/D converter as if it were in single-ended mode, and it is really in differential mode, you can get very strange results – possibly results that look valid but are really meaningless, because you might be subtracting off the signal from an unconnected input which might be varying unpredictably. Conclusion: as usual, read the fine manual, and check to see if the A/D system is set to the mode that you want.

Arcana

A/D vs. sample-and-hold. Usually these are integrated into one system and you don't need to know about it. But, just in case, you can't use a raw A/D converter by itself, you need a specialized circuit that samples the input voltage and holds it steady for some time while the A/D converter proper does its conversion.

A/D and other digital computer systems can themselves create tremendous electrical noise, that can even be a function of when data acquisition started how many channels are being recorded etc. Sometimes laptops running on battery power can help because then they are not putting noise into the electrical power system.

GROUND ALL UNUSED A/D INPUTS. This will prevent your recording

110

from an unconnected input that is capacitively coupled to a real input and getting weird results.
ALWAYS RUN AN EXPERIMENT ON A PIECE OF TEST EQUIPMENT THAT WILL GIVE YOU A KNOWN RESPONSE.

Writing computer code: beyond the scope of this manuscript, but there is one rule: THERE IS NO SUCH THING AS A TRIVAL CHANGE IN A COMPUTER PROGRAM. ALWAYS RE-VALIDATE YOUR CODE AFTER ALL CHANGES.

D/A converters: the opposite of an A/D converter. Takes a binary number and turns it into an analog voltage. Similar limits on resolution and speed apply. A particular D/A might be rated at 10 bits resolution and a maximum of 10 kHz output rate. The output is natively a 'stair case' in appearance; you need to low-pass filter the signal to get a smooth voltage.

Digital control bits: used to do something like turn a heater on and off, trigger a camera, etc. Often labeled as "TTL compatible", which just means that 0 volts = logical zero (possibly "off"), and +5 volts = logical high (possibly "on"). The name "TTL" comes from an old style of digital logic that used 0 and +5 volts as its logic standards: this style of circuitry is not as widely used any more but the voltage reference is still commonly found.

Optical isolation: often used for patient safety, especially for commercial medical equipment, and can solve many grounding problems, but at the cost of extra complexity and often these devices add considerable noise to the signal. But they can solve many problems

Some A/D systems have optical isolators built in. (Some technologies use induction instead of light but the principle is the same thing: so these are sometimes just called "electrical isolators"). Basically, these devices take an electrical signal, turn it into light (or magnetism), the light passes over an insulated gap, and then is received by a photodetector and turned back into an electrical signal. This makes it possible to transmit a signal between two pieces of equipment without have a direct wire connection between them. Sometimes this is referred to as "galvanic isolation."

Computerized processing of biological signals: once a signal has been digitized, then any mathematical operation you can imagine can be performed on it and much of what used to be done with analog filters and other processing elements is now done digitally. The binary code prevents noise from accumulating at succeeding processing stages. Anything you can imagine you can program in. The caveat is that if you did not digitize the signal at the correct resolution or speed, or there was interference or aliasing, you generally can't make a silk purse out of a sow's ear. Also: garbage-in garbage-out, it's easy to write a program that does nonsense, or works for some data sets and not others, etc.

Always run test patterns and validate what you are doing.

Also: digital filters are far more flexible and powerful than analog ones. You need analog anti-aliasing and anti-interference filters before the signal is digitized, but then the sophisticated filtering is increasingly done in a computer. Again, the warnings about getting lost with complicated programs you don't understand apply. Also some fine points: causal vs. non-causal filtering! It MIGHT appear that something happened before a stimulus was applied but this could be an artifact of digital filtering.

Know your system. Run test patterns!

Index